Date: 9/28/18

638.1 FLO
Flottum, Kim,
In business with bees : how
to expand, sell, and market

In Business with Bees

In Business with Bees

with Bees

How to Expand, Sell, and Market Honey Bee Products and Services
including Pollination, Bees and Queens, Beeswax, Honey, and More

KIM FLOTTUM

Brimming with creative inspiration, how-to projects, and useful information to enrich your everyday life, Quarto Knows is a favorite destination for those pursuing their interests and passions. Visit our site and dig deeper with our books into your area of interest: Quarto Creates, Quarto Cooks, Quarto Homes, Quarto Lives, Quarto Drives, Quarto Explores, Quarto Gifts, or Quarto Kids.

First Published in 2018 by Quarry Books, an imprint of The Quarto Group, 100 Cummings Center, Suite 265-D, Beverly, MA 01915, USA. T (978) 282-9590 F (978) 283-2742 QuartoKnows.com

Quarry Books titles are also available at discount for retail, wholesale, promotional, and bulk purchase. For details, contact the Special Sales Manager by email at specialsales@quarto.com or by mail at The Quarto Group, Attn: Special Sales Manager, 401 Second Avenue North, Suite 310, Minneapolis, MN 55401, USA.

10 9 8 7 6 5 4 3 2 1

ISBN: 978-1-63159-459-5
Digital edition published in 2018
Some of the content that appears in this book was originally published in *Better Beekeeping* (Quarry Books, 2011) by Kim Flottum.
Library of Congress Cataloging-in-Publication Data available

Design: Megan Jones Design
Photography: Courtesy of the author, except Jennifer Berry, page 19, and Ray Markim, page 125.
Cover Images: (clockwise from top left) Shutterstock, Kim Flottum, Kim Flottum, Shutterstock, and Ray Marklin
Page Layout: Megan Jones Design

Printed in China

I've had the good fortune to dedicate several books over the years to the good people who have made writing them possible—friends, beekeepers, the bees, editors, mentors, and others. And in the process, they have all deserved more credit than I can give with these simple words. So to all of you, again, thank you for the support, advice, patience, and goodwill you have all shared this time. But for all of these times, all of these tasks, all of these struggles, from the first handwritten offering to this ultimate goal of beekeepers everywhere, my life mate Kathy has been at my side. We get to work together every day, and we spend almost all of our time together every day. That is my greatest good fortune. That has made all of this possible. Thanks, Kath.

Contents

Introduction:

Transforming Your Sideline Operation into a Real Business

Not so long ago, having a hive or two of honey bees in the backyard seemed like a good idea. Maybe you moved into a new house, so your garden and two apple trees moved, too. Suddenly, it seemed your yard—and your life—needed bees.

You read all about bees being in trouble, and you heard that the world needed more beekeepers. And if you were a beekeeper, of course, you'd have honey and goodies that came from a beehive for your family to enjoy. Those would make great gifts for friends and neighbors. This could work. You could do this.

Maybe the guy at the garden store told you about a local beekeeping class on how to get started. You got some equipment from the store, some bees, and some training. You learned you would have more apples and a better garden, and you found out you could make money selling honey, too.

You took the class, read the book, and participated in a field day. You watched the instructor light the smoker, open the hive, and move frames. She pointed out the queen, brood, drones, workers, honey, pollen, foundation not yet drawn out, burr comb, and propolis. You went home smelling like smoke and thinking, "yeah, I can do this."

Then you opened your hive, and two things happened. First, you were hit with the aroma, a wonderful, beautiful mixture of bees, honey, wax, pollen, wood, even a little smoke. It was a complex, alive combination that you will never forget. And you heard the sound. It was quiet, soothing, comfortable, and easy to listen to. This was an incredible moment. Some folks go to a hive and smell the honey. Others, they just get stung. You were smelling the honey. You were hooked.

Fast forward a couple years, and you've gone from two to five colonies. The second year, there was honey. Not much honey, perhaps, but you had to figure out all the ins and outs: removing it; uncapping; borrowing the club's little extractor; then jar and label and sharing with friends and family. Then there were splits and swarms. Pretty soon, you had just over a dozen hives.

Slow at first, then more.

The first time ever I saw a frame...

Your first extractor was too small. Your second was too big. You got a big solar wax melter and went from kitchen knife to cold knife to hot knife pretty quickly. Soon your backyard was too small for all of this, so you found another location not too far away and put some colonies there. After a bit, you learned to grow your own queens, and your IPM Varroa control gave you splits to sell in the spring. A friend offered some pollination work not too far away because he couldn't keep up with it. Then the two of you started buying equipment in bulk for a lower price and selling some to pay for yours.

After you couldn't sell or give away any more honey at work, you set up a self-serve stand inside your garage door that was protected and safe. Then you sold honey at weekly farmers' markets in the summer. You discovered that retail makes money but takes time. So when some of the folks in the club began looking to buy local honey, you started selling pails; even though you made less, you invested essentially no time in the sale, which was easier. You were measuring the tradeoff of time versus level of sales.

After a while, you figured out how to balance your time between bees and your day job. Work was steady money, health insurance, and still okay. Bees were good and so was the money. And doing anything with bees was more rewarding than anything else you were doing. You were getting pretty good at making the other products. Honey was a snap, all the wax stuff was good, and you turned down pollination offers every spring. It was comfortable. But now your day job is getting in the way of the time you

Farmers' markets worked, but it took lots of time.

really want to spend with your beekeeping business. It's time to take the next step and make your bees your full-time job.

So that's what this book is about—getting from a comfortable, but unexciting, point A to a point B where you have an independent business. You know what you need to know about beekeeping—this book isn't about that. By this point, you may have 100 or so colonies, you know how much work they take,

and you know how much income they can bring in. Your percentage losses each year are in the low 20s, so replacements from splits are still free with some to spare. You know how to make the products, you probably have some experience in marketing, and you deal pretty well with both good and bad customers. All you need to figure out is exactly what your options are for an expanded business, and then how to make it happen. Perhaps you'll be expanding your

A self-serve honey stand saves a lot of time.

existing sideline, opening a retail store, reselling other people's honey, or offering education or other professional services. How serious do you want to get with all this, anyway? Let's take a look at what your options are, how to position your new business, how to expand it, and how to promote it.

1

Growing Your Operation

The first thing you'll need as you get started with the beekeeping business of your dreams is a solid plan. Start putting together a basic business plan, and take the time to be thorough about each section. It should include

- Your business name;

- A description of your operation in 250 words or less;

- A dozen or so photos, even if they are of empty fields or to-be-repurposed buildings;

- A description of your inventory over the seasons;

- Who your competition is;

- Who your customers are;

- What your labor will be, including how many people will be working and, if they are seasonal, their schedules;

- Who will be running the place when you are there and when you are not;

- Your business structure;

- Your attorney and your accountant (who must have agriculture, farm, and government experience);

- Profit and loss (P&L) statements based on projections for the first three—ideally, five—years (see the samples below);

- When and at what income you will break even and how much you will need to get started; and

- The worst things that could happen to your business—and be honest!—and what you would do if they happened.

When you have a solid plan, find somebody who works with business start-ups, but who has no clue about bees, agriculture, growing things, or beekeeping, and convince them that this is a good idea. They know what to do, and when you've convinced them you are on the right track, you're almost home.

◄ Package day is in and out, without a lot of support.

Your Competition: What Else Is Out There?

You probably already know all the bee-related businesses within a day's drive of where you want to start your operation. I use three levels of involvement to categorize these businesses; all of them can, and will be, competition at some level.

The lowest level of supply businesses are the farm stores and co-ops that just sell preboxed kits and individual pieces. They'll have the basic equipment beekeepers need to get started. They may even arrange to have a truck with packages arrive once in the spring. Because they deal with farmers, their store hours are usually good for beekeepers. What they don't have is anybody on staff who can answer questions about bees. They may have a place to call— usually the company that produces the supplies they sell—but nobody local. There's probably one of these almost in your backyard. My experience has been that they

Boxed and ready to go. Who do you ask for help?

Most of the basics, but not a large selection.

Buying some items in bulk means there's extra to sell.

don't do much or last long in the beekeeping business, but be aware of them.

Next are those beekeepers who are, essentially, very large sideliners or even small commercial outfits, kind of like you. Their first obligation is their business. They make honey, raise queens, make nucs, pollinate, and have part of a building set up with some equipment for sale. Lots of it is what they can buy in bulk and either use themselves or resell. They don't like to tie up a lot of money in inventory, but they can order almost anything if a customer asks. Their store hours are erratic, if mostly predictable, but seasonal responsibilities can—and usually do—get in the way of helping customers. There will be nucs, pack-

ages, and queens for sale seasonally, and they will generally have someone who can answer questions. Depending on where you are, there may be one small or medium commercial outfit like this close by, or even three or four if you happen to live in beekeeping country.

The third level of supply business includes the actual beekeeping supply stores. They sell anything and everything for the beekeeper. Open year round, they have lots of inventory. They're probably a dealer for two or more manufacturers (or they might even be a branch of one). Some of their equipment might be made locally. They might have a commercial beekeeper onsite, though probably not, but they'll

Store hours are short.

Not big enough for pallets, they still haul extra packages.

Hive covers.

Foundation sheets.

definitely have some bees and provide programs for clubs. Packages and nucs are available seasonally, and are usually a bit less expensive than the one-load package suppliers. Queens are available almost all year. Hours are beekeeper friendly. Prices are often more than catalogue prices, but you don't have to pay shipping, so it works out.

Spend a little time finding all of these outlets in your area. If you haven't already, visit them all and find out what they do and don't do. Ask every beekeeper you know about every business you can find. What's good, what's not, and what's just average? Be sure to ask about convenience, in terms of both hours and access. Could you do it better, faster, cheaper, or easier? What can you do that these businesses can't or won't? Better, what do you see coming down the road that the others aren't seeing or won't see? What service, action, class, or official position at an association can you offer that they aren't? Most importantly, how did the beekeepers you talked to find out about the outlets they mentioned? Who are they missing, and how well does each outlet's promotion work? Know that or go home.

Shrink-wrapped frames.

Books: basic and advanced.

Who Is Your Customer?

Once you have evaluated your competition, you have to evaluate yourself and the business you are considering. And it's your customers who will define much of it for you. Will you serve anybody who even thinks about honeybees, from the occasional consumer of a jar of honey to the smallish queen producer down the road? Will you stick to mostly small-scale beginners and hobbyists? Will you work with larger operations that may actually buy by the pallet once in a while, but mostly by filling a truck or trailer one or two times a year? Commercial operations who buy only by the pallet will be far fewer; they take far less time but far more planning, certainly.

Deciding who your customers are before you begin will answer a lot of questions. What facilities do you have or will you build to accommodate serving them? Will you need new buildings for a store and storage, or do you have existing buildings you can repurpose? Take a big enough bite that any expansion plans won't be limited, but not so big a bite that you can't possibly do it all. Both are good precursors for failure down the road, so consider the risks seriously. But *always* plan for growth. Look again at the operations that you visited and decide what they are doing well, not so well, and not at all for their customers (and your future customers). What can you do better, differently, or maybe not all so you don't do it poorly?

What Could Go Wrong?

Have an exit strategy. What will you do when the next new thing like *Varroa* comes along and you lose almost all of your customers (and maybe all of your bees) in a single season because they lost most of their bees? What will you do when the price of honey drops to way below local production costs, and most of your big customers either go out of business or radically change business direction? Or, what else could happen? Agriculture is cyclical, depending on weather, markets, government programs, and luck. That, of course, is where diversity of revenue in your

What happens when the price of honey drops below your production cost?

business comes in. Consider agritourism, other products to make and sell, and more. When one plan goes off the road, the rest of what you do should be able to cover it for you.

Financial Planning

At the beginning of this adventure, you have three tasks. The first is to determine how much money you will need to live. How much do you spend every day, week, month, and year to be able to live the way you want to live? Add up how much you spend on your mortgage, food, gas, credit card bills, kid's clothes, and all the rest of your day-to-day stuff. Next, determine how you will get that much money every day, week, month, and year. What can you do that will make that much consistently, or, if inconsistently, before you need it? The third task, once you have the basics figured out, is to ask yourself where you want to go with this operation. Simple, right?

Set aside a nest egg before you begin. Sock away at least a year's worth of income just for this. Two years' worth is better, and good accountants will tell you three because it will take that long before you have enough profit to do more than cover your costs. If you can, set a goal to not pay somebody interest on a loan. You'll hear this again and again.

Research and Resources

If you haven't already, take a look at several business plans. They'll give you an idea of what more you'll need to know. And yes, you'll need to know more than you do now. Maybe it won't be much more, if the up-scaling you intend isn't going to amount to much—just one more farmers' market, for example, or another store for honey, or adding more bees. But even if it doesn't seem like much, looking at what others have done when they started to think about getting bigger can only help.

You can search the internet for examples, and there are thousands of them. Everybody's plan is a bit different due to the nature of their business, their particular skill set, the financial advisor they had, or just because. These differences have value for you, however; they each shine a different light on the business of business, and by looking at several, you may find just the right one for you.

Some websites offer business plan models for specific types of businesses. And the good folks offering these bits of wisdom aren't doing it for free, mostly. Their business is selling you information that can actually—or maybe not—make your business adventure easier. So take their advice, but with caution.

Where to go for good information you can trust? You can start with the USDA and other government branches that provide you with exactly what you are looking for. They all serve farmers, and they absolutely want you to succeed, because the more farmers there are, the better job security they have. And, if I may, the better off we all are, because we all need farmers—and, of course, beekeepers.

Here's an overview of what you'll find when you go looking. Some of it is kind of buried in government resources, but most is easy to locate. A whole world of good information is available to help you make decisions, source what you need, and even communicate better with your family once this gets off the ground.

 Consider the saying "He who represents himself has a fool for a client." Abraham Lincoln said that, and it is as true today as it was then. Although you may have a lot of experience in accounting, local politics, or law enforcement, you most likely don't have experience in every skill set you will need to do this right the first time. As I said earlier—hire an expert.

CHOOSING A NAME

You probably already have a name on your honey label, but is it a good name for a business? If honey is all there will be, then changing or adding to it wouldn't make sense. But if suddenly you have a store and you're selling supplies to beekeepers and not just honey to regular folks, does your label tell the whole story? That's what a business name should convey: who, what, and where. Make it easy to remember for people not in the business of bees. A catchy but beekeeping-centric name may be good for the folks in the local club's members, but how well do candle, honey, and mead buyers remember "Tom's Bee Supplies" as the business they want to go back to? Would "Pine Mountain Honey" be better? Choose a name wisely because it is your face to the public, the business community, the government, and the world.

Start at the USDA: https://newfarmers.usda.gov/technical-assistance-planning-your-business (or go to usda.gov and use the search function). You'll find

- Production information: conservation, field crops, forestry, fruit and nuts, livestock, pasture and rangeland, vegetables.

- Business information: credit access, insurance, taxes and legal, certifications, land use, organic, and water rights.

- Marketing information: adding value (a real eye-opener for beekeepers), commodity marketing, contracts (another eye opener), and direct marketing.

- People-skills information: employee insurance, family issues (communication and time), work-place safety, and managing employees.

Advice and help centers can make your life a lot easier at the beginning of, and throughout, your career. Some of the best include

- eExtension is an online community for Cooperative Extension; it translates the knowledge gained from research into education programs that help provide solutions to problems facing the agricultural community. Here you can find publications from the Cooperative Extension network and ask experts for their advice. It has a focus on agricultural and food law that you will find extremely useful. articles.extension.org/ag_law

- FarmAnswers.org, a production of the University of Minnesota, and funded by USDA's Beginning Farmer and Rancher Development Grant program (BFRDP) as the national BFRDP clearinghouse, makes available several key business-planning guides. This provides a good, free starting point for business plans in the agriculture arena. farmanswers.org/Library

- BusinessUSA is a centralized, one-stop platform that allows businesses to access services and tools to help them grow and hire. It features answer services, and most importantly, free business planning tools. This is similar to FarmAnswers.org, but less agriculture-oriented. business.usa.gov/start-a-business#

- The Small Business Administration (SBA) also makes available a wealth of information about planning, launching, managing, and growing a business. It's especially good with loan information; even guaranteed loans can be found here. www.sba.gov/managing-business

- The Farm Service Agency (FSA) is a USDA website that provides basic business planning help and assistance as part of the lending process. Bring a first draft of your business plan to your appointment with the FSA for free technical assistance. FSA can connect you to local resources and organizations in your community to help you establish a successful business plan. It's also valuable when looking at cost-saving tax deductions from your beekeeping operation. Its microloan program is worth all the time it takes. www.fsa.usda.gov/Internet/FSA_File/fsa_br_01_web_booklet.pdf

- Building a Sustainable Business is a website that brings the business-planning process alive to help today's alternative and sustainable agriculture entrepreneurs transform farm-grown inspiration into profitable enterprises. Sample worksheets lend a practical perspective and illustrate how real farm families set goals, research processing alternatives, determine potential markets, and evaluate financing options. Blank worksheets help develop a detailed, lender-ready business plan or map out strategies to take advantage of new opportunities. You can find it at the FarmAnswers website: farmanswers.org/Library/Record/building_a_sustainable_business_a_guide_to_de

URBAN AGRICULTURE

The USDA offers many resources for farmers in urban spaces. For more information, check out its online toolkit, which is chock-full of information on some of the local, state, federal, and private resources available to support urban agricultural operations. The toolkit focuses on some of the challenges confronted by urban farmers, including land access, soil quality, water, capital and financing, infrastructure, market development, production strategies, and even applying for a federal grant. The toolkit also includes cost estimates for starting an urban farm. www.usda.gov/sites/default/files/documents/urban-agriculture-toolkit.pdf

Having enough room is always a challenge because bee gear takes up a lot of space, and you need a lot of space to assemble and paint it.

- AgPlan, created at the University of Minnesota, is an easy, free app designed to help develop business plans. It provides customized assistance to different types of rural businesses. Each business type has a specifically designed outline, tips or questions that help you develop each section of the plan, sample business plans, and links to additional resources. You can give access to your business plan to the reviewers of your choice, and AgPlan will facilitate interaction with them. agplan.umn.edu

- "Is Organic an Option for Me?" is a resource for you if you are interested in starting an organic business. Producing organic honey is nearly impossible in most places, but knowing the requirements has value for other products you may produce. www.ams.usda.gov/services/organic-certification/is-it-an-option

- "Tips for New Entrepreneurs," from the Small Business Administration, offers advice and other perspectives on running a business. Every now and then, it is important to take a step back and take a look at the big picture so you can dive back in with renewed energy. These are few points to revisit and reevaluate after you get up and running. www.sba.gov/blogs/tips-entrepreneurs

- The Agricultural Marketing Resource Center (AgMRC) is a national virtual resource center for value-added agricultural groups. www.agmrc.org

- "How to Start a Cooperative," partially funded by USDA's Rural Development Agency (RD), can provide technical assistance and education on starting a cooperative. www.rd.usda.gov

Choosing Your Business Structure

The SBA does a great job of explaining what format your business should take for what you want to do. Here's a short summary:

SOLE PROPRIETORSHIP

A sole proprietorship is easy to form and gives you complete control of your business. You're automatically considered to have a sole proprietorship if you do business activities but don't register as any other kind of business.

Sole proprietorships do not produce a separate business entity. This means your business assets and liabilities are not separate from your personal assets and liabilities. You can be held personally liable for the debts and obligations of the business. It can also be hard to raise money because you can't sell stock, and banks are hesitant to lend to sole proprietorships.

Sole proprietorships can be a good choice for low-risk businesses who want to test their business idea before forming a more formal business.

PARTNERSHIP

Partnerships are the simplest structure for two or more people to own a business together. There are two kinds of partnerships: limited partnerships (LP) and limited liability partnerships (LLP).

Limited partnerships have only one general partner with unlimited liability, and all other partners have limited liability. The partners with limited liability also tend to have limited control over the company, which is documented in a partnership agreement. Profits are passed through to personal tax returns, and the general partner—the partner without limited liability—must also pay self-employment taxes.

Limited liability partnerships are similar to limited partnerships, but give limited liability to every owner. An LLP protects each partner from debts against the partnership, they won't be responsible for the actions of other partners.

Decide what structure best fits your business's current and long-term goals. What products you sell will help you make that decision.

Partnerships can be a good choice for businesses with multiple owners, professional groups (like attorneys), and groups who want to test their business idea before forming a more formal business.

LIMITED LIABILITY COMPANY (LLC)

An LLC lets you take advantage of the benefits of both the corporation and partnership business structures.

LLCs protect you from personal liability in most instances, your personal assets—like your vehicle, house, and savings accounts—won't be at risk in case your LLC faces bankruptcy or lawsuits.

Profits and losses can get passed through to your personal income without facing corporate taxes. However, members of an LLC are considered self-employed and must pay self-employment tax contributions towards Medicare and Social Security.

In my experience, an LLC is the first option to look at when forming a business like this. You may already be a sole proprietor or in a partnership, but until now your cash in/cash out was fairly simple, and you most likely had insurance covering you, your equipment, and your family, anyway. And, it may still work best for you. Certainly explore the advantages before you bump your business up a notch.

CORPORATION

A corporation, sometimes called a C corp, is a legal entity that's separate from its owners. It can make a profit, be taxed, and be held legally liable. It offers the strongest protection to its owners from personal liability, but the cost to form a corporation is higher than other structures. Corporations also require more extensive record-keeping, operational processes, and reporting.

There's a lot more to corporations that you will need to explore if you decide to go this route. C Corps, S Corps, and B Corps are all possible for your business, though in all probability not at first. Explore the various forms before going this deep into your business structure.

For most small businesses registering is simple because the business model is simple.

Register Your Business

Your location and business structure determine how you'll need to register your business, so figure them out first. For most small businesses, registering your business is as simple as registering your business name with state and local governments. You may not need to register at all if you conduct business as yourself using your legal name, but then you'll miss out on personal liability protection, legal benefits, and tax benefits.

REGISTER WITH FEDERAL AGENCIES

Most businesses don't need to register with the federal government to become a legal entity, other than by simply filing for a federal tax ID number. Small businesses sometimes register with the federal government for trademark protection or tax-exempt status. If you want to trademark your business, brand, or product name, file with the United States Patent and Trademark office once you've formed your business.

REGISTER WITH STATE AGENCIES

If you become an LLC, a corporation, partnership, or nonprofit corporation, you'll probably need to register with any state where you conduct business activities.

States look at several things to determine if you are conducting business. They will consider, for instance, if you have a physical presence in the state (such as a building, land, or office), if a significant portion of your company's revenue comes from the state, and if any of your employees work in the state.

The more complex you become, the more exposure you will have and the more protection you will need.

Some states allow you to register online, some states make you file documents in person or through the mail, and most states require you to register with the Secretary of State's office, a business bureau, or a business bgency. The Small Business Association has an online list of each state's requirements.

Don't do this yourself; find a registered agent to help. A registered agent receives official papers and legal documents on behalf of your company. The registered agent must be located in the state where you register. This agent will save you time, money in the long run, and more grief than you can imagine. Hire an expert.

 The other side of the business-structure coin is don't overdo what you want to become. There's nothing wrong with being a sole proprietor as a business model. Being a corporation may be way, way more than you need. Cost is one consideration, complexity is another, and simple expediency should be another. Keep it as simple as you can, while still being protected, should be your mantra as you research the options.

If your LLC, corporation, partnership, or nonprofit corporation conducts business activities in more than one state, you might need to form your business in one state and then file for foreign qualification in other states where your business is active. Remember, your website reaches across all state lines. Your home state considers your business to be domestic, but every other state will view your business as foreign. Foreign qualification notifies the state that a foreign business is active there, and you will typically need to pay taxes and annual report fees in both your state of formation and states where you're foreign qualified. To foreign qualify, file a Certificate of Authority with the state. Many states also require a Certificate of Good Standing, and each state charges a filing fee. Hire an expert to lead you through this to avoid surprises down the road. You could simply limit your customer base to locals and avoid all this,

but imagine the disclaimer on your web page that says "[Your State] Residents Only."

For state documents and fees, the total cost to register your business will be less than $500, and the information you'll need typically includes your business name, locations, ownership, management structure, and registered agent information. To support all this, depending on the business structure you chose, you will need your articles of organization, operating agreements, or certificates of partnership if you go that route. Your agent will organize all this. Hire an expert.

Register with local agencies, if needed, but, typically you don't need to register with county or city governments to actually form your business. However, if your business is an LLC, corporation, partnership, or nonprofit corporation, you might need to file for licenses and permits from the county or city. Plus, some counties and cities may require you to register the name you're doing business as (d/b/a)—a trade name or a fictitious name—if you use one. Local governments determine registration, licensing, and permitting requirements, so visit local government websites to find out what you need to do.

You may need to file additional documentation with your state tax board or franchise tax board, typically referred to as initial reports or tax board registration. It most often needs to be filed within thirty to ninety days after you register with the state. Hire an expert.

You will also need to register with your state's Department of Revenue. Get a tax ID number, and find out what records to keep, what reports and forms to fill out, what regulations govern your business, and how to pay your state taxes. The experts you've hired really earn their money with their support now, which will be of incredible value when it comes to this part of your launch. Hire an expert every time.

What Will You Need to Get Started?

The key question to ask any creditor as you prepare your business plan is what benchmarks they look for in your peers. This can give you indicators of health—and provide guidance on what to look for in your own business plan and performance.

Business-plan work is, for the most part, tiresome and detail oriented. And, of course, the more you will depend on other people's money, the more work it will be. If all of your capital comes from you—cashing in a 401(K), using savings or an inheritance, selling the vacation home that's finally paid for—your planning can be less specific because, mostly, you are answering only to yourself. However, if the new building(s), driveway, parking, initial inventory, and/or basic initial overhead for labor, insurance, utilities, and the like will be second- or even third-party supported, you'd better have every i dotted and every t crossed. Be ready to answer every difficult question that anybody involved will come up with. And, to take advantage of what businesses can use to save money (to get available grants, to qualify for local tax breaks, and the like), you have to initially jump through all these hoops.

START-UP COST CALCULATION

Make a list of what don't you have that you'll need, and add to it as you work on your business plan. Don't forget to consider

- Buildings and access
- Down payments
- Lease, rent, or mortgage payments
- Utilities
- The ongoing costs associated with honey production and sales including bees, bottles, labels
- Vehicle
- Delivery vehicle costs
- Store equipment, including shelves, counters, money devices, internet, and phone
- Inventory
- Nontangible overhead such as the insurance needs you'll have for property
- Liability for products and people
- Professional services
- Signage
- Advertising
- Catalogues
- Website fees and production
- Fees
- Licenses
- Labor, including your salary
- Social Security and unemployment taxes
- Benefits
- Sales and income taxes
- Any items specific to new aspects of what you will be doing

INVESTIGATE SIMILAR BUSINESSES

Using association newsletters or the National Honey Board's excellent Honey Locator webpage (www.honey.com/honey-locator), find similar beekeeping-supply operations that advertise in locations far from where you are. These are businesses that you will not be competing with. Note the similarities and differences to your business, and guesstimate the size and scope of each. This might be a stretch, but visit some, if at all possible. And, because you are doing a good job preparing, look at each over the course of the year. How do they do what they do in the spring rush, the summer management period, the autumn harvest, and the winter doldrums? As long as you won't be a competitor, be bold and contact those you see as successful with questions. Include some that remain smaller, some that are mostly wholesalers or beekeepers barely in a retail business, and some that may not even have bees to speak of anymore and are mostly full-time retailers or wholesalers.

Don't overlook the basics. Find out the hours each operation is open, both daily and seasonally. What are the skills of the employees? What services do they offer outside the store? How do they handle inventory? One thing that you really need to focus on is ease of contact: How hard is it to get somebody on the phone? Do they have a useful website for basic information, ordering, announcements, or whatever it is their customers need? And don't forget the rest of their social media exposure. (See page 140 for more on social media.)

Ask questions for a bit more depth. Where do your target businesses get the goods they sell? Are they dealers for one or more of the major manufacturers? Contact a dealer and see how that works, and if they can offer any insight into the business—not so much about how much they sell, but how

business is in their part of the world. While you are at it, ask if the manufacturer offers exclusive regional coverage (and what is the size of that region?). Is credit possible, is it all cash up front, or maybe net 30, and what discounts apply, if any, for volume or frequency? Who supplies their packages? Find out how they work together from the package supplier's perspective, if you can. Do they make, or have they made, their own woodenware? Do they buy some? See if their wood fits all the other equipment on the market. You'd be surprised. The variables on wooden products are more extreme that you might expect, and it can get you in a bind in a hurry if you are relying on a single supplier.

Now look at the areas that your similar businesses are serving. What is the beekeeper population? How many are in the local association? How many associations are drawing on them for some or most of their equipment and supplies? Are there other beekeeping suppliers within a 2 hour drive of their location? Is this a prime honey-production area in the state? Are they in a major agriculture area, and what crops are grown there? Is there a lot of short-term migratory beekeeping in and out during the growing season for pollination, or longer term for a honey crop or two? This will definitely impact pest and disease transmission and how local beekeepers handle it. Does the local store handle larger-scale business or don't they deal with pallet-sized orders? If their location is mostly urban or suburban, it will realize different hours, services, kinds of supplies, and probably what customers can afford. All of this information is pretty easy to find using government statistics, state department of agriculture's apiary connections, talking to business owners, and even attending local meetings. When you find a thriving, successful beekeeping business, it has to do with the owners, the customers, the location, and how it addressed unfilled opportunities. Success is never due to luck.

The beekeeping community is a small world. Chances are you already know the folks who will be your customers and your employees.

Make a chart of the businesses you investigate. List all the attributes discussed here—plus any others that you find useful—in a column down the left side of the page. Then make a row across the top listing the businesses you evaluate. Enter their positive and negative attributes in each box and comment about each of the attributes you have listed. Could you do what they do and do you even want to do it? What don't they do? What don't they sell? Why? List everything you can about each business and compare it to your evaluation of successful, average, and barely-making-it operations. Use all of the techniques, tricks, attributes, and job descriptions from the best and brightest to begin your business plan. Look hard at what all this will cost, what you must have, what you would like to have, and what don't you need.

Over the years, I've worked with many successful beekeeping businesses and have had the opportunity to find out just why they were successful. I generally ask similar questions when visiting each so I can compare and contrast those aspects I've found to be most useful, and those to be most adverse. For example, availability of information is always useful in any of these businesses. This is a form of customer service, but business owners who know more than their customers are always more successful than those who don't. But be careful, because that immediately leads to one of the most offensive attributes I've noted: arrogance.

I grew up in a small town, working in the family grocery store. The two kinds of businesses are more alike than you might suspect at first, but of course we didn't have bees; we had beans and beef. Among the similarities is that customer service comes first, last, and always. The town we lived in was small; most of our customers were dairy farmers living not far away, meat-and-potatoes sorts of folks. One issue was

Finding employees that are as good with people as they are with bees can be a challenge.

always finances. The farmers got paid by the local creamery every other Friday, but they had to eat every day. We offered credit and the other store in town didn't. And sometimes we got the short end of the credit stick. But they were all family—what can you do?

The beekeeping community is a small world, too. More than likely, you already know many of the people who are just other beekeepers now, but will soon be your customers. And next year, and every following year, there will be new beekeepers buying the goods or services you have to sell. Certainly, if you explore cosmetics, candles, candy, or other products the rest of the world can use, you will meet people who could care less about bees but like and want their products. But mostly it's a small world out there. Your traditional, long-time honey customers will be familiar, with some new and some old customers, as long as you keep selling the same product. It's when that changes that you'll be introduced to a brand-new world with a brand-new set of rules.

POTENTIAL EMPLOYEES

What's your local labor market like? What are the skills of those you will hire? Hiring family only makes your decision a no-brainer, but if you are hiring off the street, it's a different game. What are competitive (not minimum) wages and benefits in your area? Creating a part-time position will be less expensive, but you'll lose out on that really good employee who needs a full-time job. A temp agency? Really? A retired beekeeper? Maybe. A housewife looking for a part-time job? Not likely. A seasonal employee from a nearby commercial operation? Great, if you can find and keep him or her. Somebody who wants to be where you are now? Perfect. Don't you wish you could have found that job a few years back? But then, you are training your competition, aren't you? Labor can be tough, but the right person can be one of your most valuable assets, so choose wisely.

Buy or Build?

Now that you have analyzed and identified successful businesses, talked generally with those who can assist you financially and otherwise, listed the physical requirements of what you'll need to succeed, and have a good idea of what your operation will need, here's the big question. Is what you have enough or, rather, good enough? And looking into your crystal ball, will it be good enough when you have all the stuff you need inside, all the people lined up for packages, all the seasonal stuff they need, and some they don't? If your garage just won't do for a building and your backyard isn't a parking lot, then your choices are to build to suit or to buy what's needed. Sometimes, the least troublesome road is to buy an existing bee supply business or at least an existing facility that you can turn into the business you want.

Zoning is the thing to check into. If what you want works within the regulations, go for it. But don't go until you know.

So, is that bee supply place down the road for sale or could it be? That may be the best choice. Why start from scratch when somebody else has already taken the risks, made the right choices, and burned the candle at both ends. If buying a local (or nearly local) operation is a possibility, you would do well to explore it. The first thing to find out is why it's for sale. Are the owners retiring because business is not what it used to be, beekeeping isn't what it used to be, or the competition from Amazon is more than they want to deal with? Know why before you take another step.

Is this a business you'd be interested in?

Plan on more than enough room to accommodate events and sales days.

But if you're buying, a bank or government loan might cover you long enough to become established. A combined honey crop the first year might do it. A good accountant studying the books that cover the last five years will tell you if paying the loan off in a certain amount of time at so much a month is doable. And they like to do this stuff. Explore this if it's possible.

If there isn't a bee supply business nearby that's just begging to be sold, and chances are there's not, the next best choice is to either buy a facility that exists that meets your requirements or to build one. But here's that same word of caution: why isn't there a bee supply outlet nearby already? Not enough beekeepers, a lousy honey or pollination location, too-strict zoning regulations, or what? Given that beekeeping has been growing for the last 10 years, with no sign of letting up, there might be a reason that the area doesn't have enough customers to support a small business. Find out.

But when you decide to proceed with a purchase or buying a building, there are some major requirements you have to consider. Let's take a look:

- **Parking, driveways, turn-arounds, and loading areas.** Think about package day and the day you bring in honey supers. Think about deliveries and loading trucks. And, where will you stack supers for washing after extracting? If people can't get in or have to walk a ½ mile (1 km) to buy, they won't be back.

Is covered storage good enough?

- **Bee equipment storage for what you use and what you sell.** Think about the supers, trucks, frames, tables for cleaning, covers, pallets, feeders, foundation, barrels, beeswax, pails, bottles, wood repair tools, and saw horses, all things you may only be able to fit a few of at a time inside the actual store. But consider that your warehouse may actually be your store, with everything inside in big piles and rows and stacks. That's possible and will save you a bundle.

- **Beehives.** Indoor wintering is definitely the future; can you do it with your existing buildings? Probably not, so you'll need insulation, cement floors, stable electric service, ventilation, overhead doors, refrigeration and backups, lighting, a driveway and a solid staging area right outside the overhead door.

And what about cool package storage?

What about overwintering indoors? It's the future for honey bee health.

An attractive entrance provides an appropriate welcome . .

■ **The actual store.** If you are going to be a dealer for a major manufacturer, you have to have a real store, because the big guys require it. That means a building (though it may be your warehouse), mud-free driveway, parking, and loading dock areas. Inside, there should be room for shelving, wiring for good lighting, counters, backroom storage, an office with internet, and external and internal security systems (cameras, fire detectors, and alarms). Indoor package and queen storage will most likely be part of the store, so double or overhead doors might be part of the building (see parking and driveway, page 29). If you use carts inside for anything, make sure your carts fit through your doors.

. . . and any easy to use exit will be appreciated.

You'll need counters to display products and protect things like books and small items.

Basic equipment.

How much shelving will you need?

Group like items together but always have a surprise.

A checkout counter, with internet, phone, and electric connections, is a must.

And don't forget the must-haves.

You'll need a place to assemble and store wooden equipment.

A workshop will probably be part of the deal.

- **A workshop.** If you are making some of your products, you'll need the right equipment, wood storage, an equipment assembly room, finished goods storage, maybe a painting area for those fancy supers you sell to farm stores, and doors that lumber and boxes will easily pass through.

How much warehouse space and equipment will you need, and how much do you already have?

Container storage is a viable alternative.

- **Storage for empty equipment.** That includes supers, excluders, covers, bottoms, feeders, pollen traps, anything that needs to be fixed, inner covers, frames, dirty wax, anything that needs to be cleaned, and anything else that doesn't go anywhere else. You'll also need doors that are bigger than the biggest things you, ever have to move.

- **Bees.** Can you have five, twenty, or fifty colonies out back because it's a good beeyard, it's close, and they make honey there every year? Do you want customers dealing with bees coming and going, swarms, stings, and . . . well, what about your bees?

- **A dedicated honey house.** The garage, basement, or kitchen will no longer work. They aren't big enough, aren't available often enough, and probably don't meet food processing standards in your state, which you absolutely need to find out about before building. If honey is going to be a measurable part of your business, you need a honey house that meets legal requirements and is easy to work in.

And where will empty equipment be stored?

You may find the right place to do all this for right now, but what about when your company grows, or you expand any part of the business? What if you want to make more wood gear or you need more indoor winter storage? Does the location allow for future expansion, or will you have to do this again in five or ten years somewhere else? Could one building be repurposed from one task to another without major renovations?

If you are going to build something new, you must consider growth, repurposing, downsizing, increased truck traffic, and all the rest. Also consider urban expansion down the road. You may be somewhat isolated in the beginning, but suddenly there may be a parking lot and grocery store next door, with lots of pedestrian traffic. That's okay most of the time, but package day? Probably not so good.

 There's always the possibility that the federal, state, or local government will decide a new four-lane bypass is needed in your community and suddenly you find people can't get to your place without way more trouble than it's worth. Or, part of your land comes under eminent domain for this and suddenly you have way less space than you did. Scout out both possibilities before you sign on the line.

Your Honey House

When you decide that a commercial-oriented honey house is in your future, there are a host of precautions you'll need to take. Here are some of the more obvious ones, but you need to find your specific requirements before you begin. Then find the people who will eventually sign off on your final project. Keep them in the loop from day one to avoid expensive surprises down the road. Find out, too, if there are differing requirements depending on the level of sales you have each year. And if you have to ramp up, figure that next year may be different than this year, and if making any changes, scale up rather than keeping it for what you need this year.

If you're buying new equipment, make sure it fits, it works, and it's okay to use, regulation wise.

How does the equipment you already have in a new or existing honey house stack up to any local, state, or federal food-safety regulations?

HONEY HOUSE REGULATIONS

I've borrowed from the Michigan Department of Agriculture and Rural Development (MDARD) honey house regulations (www.michigan.gov/mdard/) because they are typical, easy to understand, and complete. These regulations should be used as a guide to set up your honey-processing facility to meet food-safety standards. Contact your state's department of agriculture and local board of health for licensing information and updated standards.

Processing Facility. A honey house is any stationary or portable building, or any room or place within a building used for the purpose of extracting, processing and/or handling of honey. It must be constructed in a sanitary manner, suitable to the operation, well maintained, and in good repair and vermin proof. During the processing of honey, this place is to be used only for operation or storage of equipment incidental to the activity. (This room excludes the family kitchen.) The grounds adjacent to this building should be free from all litter, waste, refuse, or anything that may constitute an attractant, breeding place, or harborage for rodents, insects, and other pests.

Floors. Floors of all rooms in which honey is handled must be of impervious and easily cleaned material and must be smooth, in good repair and kept clean. Floors that have a drain should drain into an approved sewage disposal system.

Walls and Ceilings. Walls and ceilings shall have washable surfaces, be clean, and kept in good repair.

Ventilated Openings. There shall be adequate ventilation, and all ventilated openings shall be screened to exclude insects. It may be that honey bees enter the extracting facility, but efforts should be made to control entry and to relocate bees once inside.

Lighting. A permanently fixed light source shall be sufficient to permit efficient operations. All lights over the processing area shall have protective shielding or be shatter resistant.

Have big enough doors.

Water Supply. There shall be an adequate source of safe potable water under pressure in accordance with local sanitary codes. Water from a nonpublic source (private well) shall be tested at least once a year. Contact your local health department about testing procedures and locations.

Approved Septic Systems—Water Disposal. MDARD does not approve septic systems but does require that the system be approved by the local health department. Beekeepers starting up a facility should contact their local health department for information on the approval process.

Toilets. Toilet facilities, including hand-washing basins, shall be conveniently available to honey-house personnel.

Vector Control. No chemicals or poisons shall be used to control insects or rodents while supers of honey are stored in the facility or extracting of honey is proceeding. Follow label instructions and use only approved pesticides.

Construction, Care, Use, and Repair of Honey House, Containers and Equipment. During operation, the honey house shall be used exclusively for extraction, processing, packing, or other handling of honey and storage of equipment related to the activity of the honey house. Containers shall be free of

Is this old wax sump going to work in your brave new world?

This has to go!

internal rust, cleaned before use. Honey shall not be packed in containers that have previously contained pesticides, creosote, petroleum products, paint, toxic substances, or anything that would react with honey. All equipment should be covered when not in use. Extracting equipment shall have hard surfaces that will not permit reaction of the acid in honey and should be lubricated only with food-grade grease.

Heating Equipment. No boiler, oil stove, or other heating equipment that gives off odor or dust may be used within the honey house, unless it is properly ventilated and complies with fire regulations.

Workers Sanitation. Workers shall maintain a high degree of personal cleanliness and wear clean, washable outer garments.

Hand Washing Station. This area shall be conveniently located, easily accessible, and supplied with tempered water, soap, single-service disposable paper towels or an effective sanitary drying device.

Ware Wash Sinks/Cleaning and Sanitizing. The regulation of honey facilities requires that all food-contact surfaces, including utensils and equipment, shall be cleaned as frequently as necessary to protect against contamination of food. The standard method of cleaning and sanitizing equipment and utensils is a three compartment sink. In addition to the wash sinks, the facility needs to have at least one conveniently located hand washing sink. If using fresh hot water, sanitization equipment / utensil surface temperatures must reach at least 171°F (77°C).

Do you have room for a three-compartment sink?

Processing Equipment. Processing equipment shall be constructed of durable, smooth, and easily cleanable materials, well maintained, and stored to avoid contamination.

Bee Hive Equipment. Extracting supers shall be clean and free of contamination. Medication is to be applied only when required following label directions. Supers should be covered, transported in clean vehicles and stored in a clean/sanitary area prior to extracting.

Extracting Honey. Uncapping should be done under sanitary conditions. Extracted honey should be covered if the process is not continuous. Each lot is processed separately and a sample of each lot shall be retained with proper label identification.

Straining of honey shall be done with a clean mesh-cloth material or stainless steel mesh. Reuse of cloth mesh is not advisable. If honey is filtered, the filters should be checked for damage to ensure they are clean and free from extraneous material.

Honey Bottling and Labeling Requirements. The processing of honey for bottling shall be done in a clean/sanitary devoted facility. Retail and wholesale honey containers should be labeled with product identity, ingredients if any are added, a production or lot code, name, address of manufacturer, packer or distributor, and net weight.

Cooperative Honey Houses (Central Kitchens). As a cooperative honey house, multiple processors can use the licensed facility. The processors that use the facility are not required to obtain a license. Accurate records of members using the facility will be kept. From time to time, MDARD will examine the production records to verify that honey found in commerce was actually produced at the cooperative.

Transport of Supers. Vehicles used to transport supers after removal must be clean. Supers must be covered during transport to prevent environmental contamination. Storage of supers must occur in a clean and sanitary environment.

Most states offer a no-cost plan review service to beekeepers who are starting up or adding to a honey processing facility. The plan review process can help prevent costly revisions by evaluating the proposed facility before construction begins. This process also facilitates communication between the inspector and the operator throughout the construction process and makes it easy to get questions answered. For example, they can identify often-missed gaps in the plan such as an unsealed floor, improper sinks, or lights that aren't sufficiently covered, so that you do not have to remodel or replace fixtures after they are installed.

EFFICIENCY IN SET-UP

Ever consider time/motion studies in this harvesting, processing, and returning job you have to do every fall? Still doing it the way you've always done it?

I'm going to guess that your existing honey extraction facility is not a time study–designed, efficient operation, but rather a make-it-up-as-you-grow-along collection of machines, stools, tables, pails, and whatever else you've accumulated over the years. Moreover, your space probably hasn't changed much, so you've gotten more and more crowded as you try to get more stuff in the same space. Am I right? Let's start this process over with efficiency in mind. If you could save 10 seconds on every frame you have to handle, from beeyard back to beeyard, how much time would you save in a season? Even 2 seconds on every frame can make a difference. Now imagine saving minutes off boxes. The savings in labor and increased efficiency means you can either manage more hives in the same amount of time or reduce the cost of those you manage now.

So before the concrete is even poured, design your workflow around the equipment you will have in the near future—not today—so that the movement from the loading dock through the oversized doors into the warming room and over to the uncapping stack is as easy and linear as possible. Then, think about moving the now-empty boxes to the cleaning barrel, while frames from the uncapping tank go into and out of the extractor and boxes from the cleaning

Will your insurance cover this type and amount of damage?

barrel go to the extractor, frames go into the boxes, and boxes are and moved to the outside holding area to be power-washed and dried. Double all that if you'll be using two extractors. Then think about the trip back on the truck to the beeyard or storage.

Is all this hand moving with a two-wheeler, or will it soon require a power lift of some kind? Got room for that now? Where do you stack the pallets before you start, and how do you get more into the room when you run out? Design straight shots so you aren't twisting and turning. Make sure the doors are big enough to go through at an angle and that you can push them open with the pallet the boxes are on.

How many times do you handle a box from hive back to hive? A frame? Can you reduce the number of times, or increase the number of boxes or frames each time you move some? Where will you be in 2 years? Set standards so you know if it's working, then work to improve it. Where can you cut a motion, a turn, or a second move? And then, pay your help for their efficiency—supers per hour, frames per minute, colonies per day—not their time. Get a stopwatch, and use it.

Insurance

If you're moving from a sole proprietorship or partnership to an LLC, the insurance you'll have to get now separates your personal life from your business world and you'll need that. You'll certainly want to invest in product and liability insurance, especially if now you'll have lots of people on your property with live, stinging, venomous insects a part of the daily routine. You should have had liability insurance for your honey sales, but too often small-scale beekeepers neglect that cost, and it is hardly ever an issue.

The insurance you get will also be dictated by who else has a piece of the operation. The folks who loaned you money certainly have a say, but so does anybody who has a stake in what you are doing and what you can lose if it all goes up in flames (or lawsuits). Listen, too, to the accountant you hired; that's why you hired her.

Liability insurance has two purposes. It makes payments to an injured party on your behalf, and it defends you against lawsuits brought by a third party alleging liability under the policy's coverage. Basically, something happens, damage or injury occurs, and the person responsible is legally liable. To recover damages for negligence, the injured party must show you were responsible, you dropped the ball somehow, someone was injured or something was damaged, and there is a value to recover from the damage. Or, one of your employees may cause harm or be harmed by something while performing his or her duties. There is also the possibility of harm during an agritourism event.

Of course, this is much more complicated than described here, and because of a beekeeping operation's less-than-common opportunities for harm, you need to carefully spell out what your operation will be engaging in as a business when talking to an insurance agent.

Have you thought about recall insurance, for damaged honey recalls, for example? Because the National Honey Board deals almost exclusively with packers, and just the biggest packers, they will have a handle on new rules coming into effect, or older rules changing that may affect your honey business.

You will need all the standard insurances, too: property, vehicle, and all the rest. When you are figuring startup costs, this is a fairly large factor. Be sure to consult with both farm agents and insurance agents so you have everything covered.

 Don't forget people insurance. If you are leaving a job that gave you healthcare coverage, you'll have to replace that. Perhaps a spouse's policy can do that, but if not, you'll have to find a source of coverage. Don't overlook the need to add life insurance, too.

Funding

Now come the financial questions. Can you pay for all this?

If you've been building slowly over a few years, putting what you've made back into the business for growth, you may be at a point at which you can use savings or small advances from family to start out. Starting slowly and keeping an eye on both where you are and where you want to get to is a time-tested and safe business model. Earn, learn and reinvest.

If it is going to require some additional investment, the contacts you made when researching your business plan will be invaluable. The quality of information available from the University of Minnesota's AgPlan program is especially impressive, but others, such as the USDA's Farm Service Agency (FSA), are helpful, too. The small business loan people are an absolutely indispensable resource for anything you need to know, do, fix, or make. Their microloan program has facilitated thousands of businesses like yours, so certainly investigate that.

Don't forget family members who know you better than any bank or loan office, and who are less interested in the loan than they are in seeing you satisfy a dream. They are aware of your ability and drive.

Your good accountant and tax person will be able to help you with a figure. To pay existing bills, purchase needed and new equipment, and plan for additional labor and purchases for the next 3 years, your operation must earn a certain number of dollars every hour, day, week, month, and year. If you take care of the hours, the rest fall in place, but that isn't always that easy. Not correctly calculating your startup costs will probably put you out of business sooner rather than later. You can't rely on a fifth year of an extraordinary honey crop to make up for the proceeding bad years. That's never a sure bet.

2 What to Sell— Products Made by Bees

For the purposes of this book, let's make some assumptions. First, before you started down the road to self-employment, let's assume you were already in the honey business to some degree. It probably involved some pollination, nucs, and queens, but ultimately honey was important. That's what started this in the first place. Let's assume you have steady, regular customers who come to your house or stand, you have a farmers' market routine, and maybe you supply a few local high-end restaurants or specialty food stores. You've pretty much been focusing on direct retail or high-end wholesale rather than bigger wholesale customers. But maybe you have some of those, who buy from you to resell to their customers, too. Let's assume you know how much honey you have to produce to supply those outlets, how many colonies it takes to produce that much honey, and how much the process costs. That means you know how much to sell honey for to stay in business.

Are you just growing your business while not changing the rest of your life? If keeping your day job is still in the cards, some of your costs will stay under control. You'll have a steady paycheck and health insurance. Crop failure wouldn't be the end of the world, and part-time help would still be doable. Of course, you can't buy time, so you'll have to figure that out. But this kind of careful expansion is possible as a first—maybe second—step to setting out on your own.

If honey is to remain your only product, your path is pretty straightforward. If you're expanding sales, you'll need more honey, more colonies, more input costs, more time, and more outlets. You'll need more storage for bottles, labels, and pails. You'll probably need some new (or better used) stuff, such as moving equipment, stainless to make the honey house more efficient, or maybe a bigger delivery vehicle.

There are several ways to expand an operation like this: Change or add honeys. Go from wildflower to several varietals. Include raw or infused honey. Stress local or seasonal honey. Increase your marketing horizons. You won't have to do a lot differently, actually. It's still honey you're producing, so you won't need a lot of new or different equipment, and

you probably won't need much more room. More colonies, yes. More storage, probably. More time, absolutely.

Find a way to move colonies for varietals more efficiently. This means a trailer or a truck. And once you have bees and a way to move them, doing at least some pollination, if you don't already, is a no-brainer. Besides, if you have to get a larger vehicle, you have to use it more than once a week or so to make it pay for itself. Moving bees is the way to do that.

Forms of Honey to Sell

What kind of honey will you be selling? You have lots of options:

- Local
- Blossom
- Honeydew
- Artificial (yes, a type)
- Extracted
- Creamed
- Comb
- Chunk
- Unfiltered (and unheated so it's raw)
- Strained (extraneous removed; pollen remains)
- Filtered (nothing is left; no pollen)
- Varietal
- Infused
- And yes, simply honey (but which grade?)

These are all moving targets at the moment, as definitions and food safety regulations are evolving both in the U.S. market and on a global scale. So, to keep things well defined, let's look at each of these briefly.

LOCAL HONEY

If you are using the term "local honey," you need to have an easy-to-understand definition posted somewhere obvious that spells out exactly what you mean. Is it local if it's produced in your state? Your county? Within 100 miles or 10 miles? Customers want it to be local because of the myth that local honey helps seasonal allergies. But whether it does or not, local honey supports local beekeepers, so promote that. Rather than not advertising "local," you can refine the concept by putting on the label that it is "Produced in [your state] by a [your state] beekeeper." For example:

> *Autumn Glory Honey*
> *Produced by Tall Pine Apiaries in Medina, Ohio*

You can put an additional label on the back or top that says "Local Ohio Honey" to reinforce the location.

BLOSSOM, HONEYDEW, AND ARTIFICIAL HONEY

The definition of honey has always been, well, fuzzy. As I'm writing this, there may be, finally, a more precise definition. You can't prove something isn't honey until you know what it is, first. To get around this fuzziness, those who enforce food safety laws have come up with a different set of standards. Rather than say what honey is generally, they've defined several products that are often referred to as kinds of honey while these definitions have been up for discus-

Sourwood is but one of many exotic varietals you might try.

sion for a long time, it's the best we can do for the moment.

From the USDA and FDA, and found at www.ams.usda.gov/grades-standards /extracted-honey-grades-and-standards

6.1.1 Honey. Honey is the sweet substance produced by honey bees (*Apis mellifera*) from the nectar of plants or from secretions of living parts of plants or excretions of plant-sucking insects on the living part of plants, which the bees collect, transform by combining with specified substances of their own, deposit, dehydrate, store, and leave in the honeycombs to ripen and mature.

And…

6.3 Honey composition. Honey consists essentially of different sugars, primarily fructose and glucose, additional carbohydrates including maltose, sucrose, and other sugars. Additional substances naturally derived from the collection of nectar by honeybees include proteins, amino acids, vitamins, and minerals.

6.1.1.1 Blossom or nectar honey. Honey that comes from the nectars of plants.

This is simple wildflower honey. Nothing varietal, perhaps not local—what a shame. But considering the source, perhaps that's as good as you can get. If you're stuck selling this product, get rid of the "wildflower" in the name and give it something unique: Spring Tree, Autumn Blend, or Summer Wonder,

Comb honey. Have parts, labels, and foundation on hand.

for instance. Or give it a name that promotes a use, like Bar B Que's Best, Me and Tea, or Feel Better Soon. Or give it a location, even if you're not sure where it's from, like West Coast Wonder or Southern Pride. All of these give the buyer a sense of possession and a name to ask for again if they really like it.

6.1.1.2 Honeydew honey.

Honey that comes mainly from excretions of plant sucking insects (*Hemiptera*) on the living parts of plants or secretions of living parts of plants.

This isn't commonly sold in the U.S., but it is occasionally produced here and then marketed overseas where it is not only common, but desired. When sucking insects, such as aphids or plant bugs, attack certain plants, they pierce the plant tissue and suck the sap they encounter. Their bodies process it, and then the excess is excreted as waste, which has a high sugar content. Bees collect this and process it as nectar. Generally, it is dark, and it can be strong, with a distinctive flavor not well appreciated by U.S. consumers.

6.1.2 Artificial honey.

Honey composed of mixtures based on sucrose, glucose, or invert sugar, generally flavored or colored and prepared to imitate natural honey.

6.1.3 Adulterated honey.

Honey that has been mixed with additional substances such as, but not limited to sucrose, commercial glucose, starch, gelatin, water, etc. In accordance with Title 21—Food and Drugs, Chapter 9—Federal Food, Drug and Cosmetic (FD&C) Act, Subchapter IV—Food, United States Code (USC) § 342: "Adulterated food. A food shall be deemed to be adulterated—. . . (b) Absence, substitution, or addition of constituents (1) If any valuable constituent has been in whole or in part omitted or abstracted therefrom; or (2) if any substance has been substituted wholly or in part therefor; or (3) if damage or inferiority has been concealed in any manner; or (4) if any substance has been added thereto or mixed or packed therewith so as to increase its bulk or weight, or reduce its quality or strength, or make it appear better or of greater value than it is."

LIQUID AND CREAMED HONEY

6.1.4 Extracted (liquid) honey.

Honey that has been separated from the comb by centrifugal force, gravity, straining, or by other means, including crushing.

6.1.5 Granulated (creamed) honey.

Honey made by blending one part finely granulated honey with nine parts liquid honey. The mixture is stored at about 14°C (57°F) until it becomes firm.

Finely granulated honey is often called a starter. The finer the crystals in the starter, the finer and smoother the finished product. Honey left to crystalize on its own often has large, coarse, uneven crystals. And it often crystalizes unevenly, leaving some very dilute honey, which can later ferment, in the mixture.

COMB HONEY AND CHUNK HONEY

6.1.6 Honeycomb (comb in wooden or plastic frame) honey.

Honey direct from the honey-filled beeswax comb as stored naturally by the bees in the hive.

Comb honey can be sold as an entire frame of capped honey; as pieces of comb honey removed from the frame, drained, and placed in containers; or produced by the bees directly in the container it will eventually be sold in.

6.1.7 Chunk (bulk) honey.

Comb honey in a container with liquid honey poured around it.

If chunk honey and comb honey are relatively unfamiliar to your customers set, start with smaller containers so you can keep costs down and curiosity up. Price can be a barrier for a new product. If these honeys are new, feature them in your online ads and highlight them in your store.

Make sure you have a steady, secure supply of your chunk and/or comb honey, because if you create a demand that you can't meet, your credibility goes out the door, along with your potential customers. If maintaining a steady supply from a local beekeeper, whether yourself or someone else, is an issue, you can approach sales two ways. First, you could treat it like seasonal vegetables. Get your customers excited

Put honey dispensers right next to the honey.

before harvest, then with great fanfare and flair, announce it is available for oh, so short a time. Let them know it will be gone soon, so they should hurry and get the best local raw honey that can be had. Or, always have a steady supply of exotic comb and chunk honeys available. Stock orange blossom, sage, alfalfa, and tupelo, for example, all from faraway places and sources not locally available. This makes you the go-to supplier for those interested in something other than mundane local honeys.

Of course you can do both, depending on your customers' experience and your ability to carry that level of inventory. But be warned that unsold, granulated comb honey can quickly become a costly burden.

RAW HONEY

6.1.8 Unfiltered/unstrained (raw) honey. Honey that has not been filtered or strained, as described by the U.S. Standards for Grades of Extracted Honey and may include extracted or non extracted honey and whereas most of the fine particles, pollen grains, air bubbles, comb, propolis and other defects normally found in suspension may be present.

Raw honey is as tricky to define as local honey. "Unheated and unfiltered," without further explanation, is usually how it's described. But honey in a beehive can be as warm as nearly 90°F (32°C). Sometimes, it gets even warmer in the hotter parts of the country. If you remove honey from the hive and it is hive temperature, has it been heated? It's certainly warmer than room temperature. If you harvest it, let it sit, put it in a warming room, and raise it back to hive temperature, has it been heated? See how silly this can get?

I suggest using this compromise: "This honey has not been artificially warmed more than the normal temperatures found inside a healthy beehive." And leave it at that.

UNFILTERED VS. STRAINED VS. FILTERED HONEY

6.1.9 Filtered (processed) honey. Honey of any type defined in the U.S. Standards for Grades of Extracted Honey that has been filtered to the extent that all or most of the fine particles, pollen grains, air bubbles, comb, propolis, and other irregularities normally found in suspension have been removed.

6.1.10 Strained (processed) honey. Honey of any type defined in the U.S. Standards for Grades of Extracted Honey that has been strained to the extent that most of the particles, including comb, propolis, or other irregularities normally found in honey have been removed. Grains of pollen, small air bubbles, and very fine particles would not normally be removed.

Is honey still honey if it doesn't have pollen in it? Some people feel that if you strain honey so finely that the pollen is gone, much of the rest of what makes honey actually honey will be gone, too.

Do you leave all the bee parts, wood splinters, and chunks of wax in the mix and bottle them all up? That results in honey as unfiltered as you can get. Or do you just get the biggest, grossest pieces out, leaving smaller pieces of wax, certainly pollen, and perhaps some of the other minutiae? That makes strained, but not filtered, honey. Generally, the filtering process includes heating the honey to between 100 and 150°F (38 and 66°C), then pumping it through at least one filter/strainer/ceramic sieve that removes almost everything but the sugar, including the pollen.

Here's a heads-up for when you're straining honey. No matter the filter you use—nylon cloth, strainer from the kitchen, double wire filter, paint filter, or milk filter—once honey starts running through it, some of the holes will become smaller as they fill up. If the strainer isn't cleaned routinely—and the necessary frequency varies—more and more of the debris in the honey will be held back. You'll essentially be heading toward a filtered product. So be careful what you call it and how you treat it.

KNOW YOUR HONEY

Selling honey to beekeepers? Really!

You would be surprised how much honey you can sell in a store that caters to beekeepers. Beekeepers like honey, and they especially like honey they don't have. Your best customers, however, will be people within medium driving distance who absolutely have to have what they consider local honey. And, because you know every local beekeeper, you will have more types of local honey than anybody around. Don't overlook this staple for your outlet. The more customer-friendly your access is, the more product you will be able to move. Even though you are unique, you are still competing with regular grocery stores and farmers' markets for hours, parking, and selection. A customer must want to come to your location. Make it worth their while.

An excellent way to do that is to know more about honey than your customers do. If you order orange blossom honey, how do you know it is orange blossom honey? Because it says so on the label? Good luck with that. Know your source, and know what orange blossom honey tastes like. This is basic stuff, but you'd be surprised how many beekeepers don't know squat about honey. They know how to make, harvest, extract, bottle, and sell it, but they don't know about honey. Learn. Attend classes that teach tasting. Learn pairings. Read books and watch videos. A beekeeper who knows less than a customer about something as fundamental as honey shouldn't be selling it.

Display different kinds of honey in different locations.

Don't forget honey sticks.

HONEY VARIETALS

6.2.1 Name of the food. In accordance with the FDA's honey-specific Compliance Policy Guidance (CPG) 515.300: "A honey may be labeled with the name of the plant or blossom provided that the particular plant or blossom is the chief floral source of the honey, such as 'Orange Blossom Honey' or 'Clover Honey,' and provided that the honey producer is in a position to demonstrate that the plant or blossom designated on the label constitutes the chief floral source of the honey." In the case of imports, labeling must be deemed informative and truthful by FDA in compliance with their Food Import Program requirements.

Simply calling something blueberry honey doesn't mean much until you define and prove, if necessary, what's in the jar. There are various definitions that allow you to call a honey a varietal or monofloral. With one definition, at least 51% of the honey must be derived from the source stated on the label. That means that 49% can be something else. Or you could say the predominant source of the honey in a jar is a particular varietal. That means the honey can be 30% the flavor on the label, but contain about 15% each of five different other kinds. That 30% is predominant. Is 30% blueberry honey really blueberry honey? How are your ethics? And what about your supplier's?

If you are going to produce or sell a varietal honey, choose a good definition and be able to back it up. This can be expensive, but it's worth the cost. Reputable packers will have the honeys they purchase examined for several things, variety (actually pollen numbers) being one of them. The pollens found during pollen examination can confirm the source country of the honey. Pollen examination is a pretty exact science, and you can't do it in the back room with a $100 microscope. If you are going to sell varietal honeys, require proof from the seller and yourself. Be right, don't guess, and charge a premium price for the product.

HONEY STICKS

Honey sticks are golden. If there isn't one near you already, become a regional distributor or manufacturer. Be the source for every association that sells honey sticks at every event they sponsor. The upfront cost is reasonable, you don't need a huge inventory at any time, and the markup is also reasonable. You can get generic brands from the supplier, or you can use your own or your customers' honey for a more personal touch. These cost more, but they are worth more. You will sell honey sticks to every retail outlet that sells your honey, and it's another income source you can add with little investment. There are machines available for making your own, too, so explore them.

ORGANIC HONEY AND CERTIFIED NATURALLY GROWN HONEY

Though attempting to keep to either organic or Certified Naturally Grown standards is admirable, producing organic honey, essentially, isn't possible almost anywhere on earth except Brazil, which, for some customers, may be acceptable. Depending on your clientele, buying bulk Brazilian honey may be reasonable if there is demand and you can be the only local source. The trouble is, Brazil's organic honey, which is produced in the non agricultural jungles of that country, tend to be dark and relatively strong. Most of your customers, meanwhile, will want traditional light and mild honey. But, if there is demand…

A Certified Naturally Grown designation has a lot of value for some agricultural crops, but it hasn't taken off like I hoped it would for honey, at least not yet. It's a fair amount of work to get things in line to qualify, and more to keep it that way, without a lot of recognition from the public. A not uncommon question is, "You grow honey?" and then that has to be explained. You don't "grow" the honey, but you follow organic guidelines in health care, harvesting, and processing of your honey without the honey qualifying as organic If it's important to some num-ber of your customers, you may choose to check into it, and you will be a better beekeeper if you follow the guidelines they require. It's also one more thing you can brag about.

INFUSED HONEY

Finally, there's infused honey. Ever made any of that? You can use citrus peel, peppers, cinnamon, or almost anything that will add flavor to your honey. Note that because you no longer have "honey" in that bottle, label and bottling regulations may come into play. Farmers' market sales are good for this, as are CSA sales and home store sales, but a grocery store that requires stricter labeling requirements might fall under different guidelines. Find out first.

Pepper-infused honey is popular and easy to make. If you haven't before, try this and see.

Gather 1 pound (0.5 kg) or so of fresh hot peppers. Jalapeños, habaneros, tabascos, and chilies all work; choose hotter or milder to taste. My friend Buzz uses what he can get from a neighbor who grows veggies for a living. He fills a 300-pound (135 kg) warming and bottling tank with honey. Then, he takes the fresh peppers and, using a blender, chops them up until they aren't quite pureed, but close. He puts the mixture in a nylon bag made from the same material he uses to strain his honey, ties it

Farmers' markets and infused honeys are a natural . . .

. . . right next to the fruit used to infuse the honey.

shut, and puts it in the tank full of warmed honey. He turns the heat off and lets it sit for a week or so. Then he warms the honey again to slightly above room temperature, lets it cool, and lets it sit for another week. Then he removes the bag, squeezes any juice back into the tank, discards the remains, and thoroughly mixes the honey. You can regulate the pepper heat of the final product by how many times you do this, how many peppers you use, and how long you keep soaking the bag. He produces "Hot," "Very Hot," and "Ouch" honeys. His label says "Hot Pepper Infused Honey," and the list of ingredients says "Honey" and "Jalapeño Pepper Infusion." He sells it only in places where he can let people taste it so they know what they are getting—and just how hot "Ouch" is. He sells a lot of them.

You can mix any spice you want with your infused honey. Fresh spices work better, faster, and stronger than dried. If you use fresh herbs or citrus peels, watch for increasing the moisture content of the honey up to fermenting levels. But soaking them until the honey has pulled all the moisture out of the leaves is the way to go. And the stronger the herb, the stronger the flavor, and the less moisture you will have to deal with.

And again, check your local and/or state labeling requirements before going full public with this.

HOW TO DECIDE

If you haven't already, find out what a casual honey buyer can purchase from local stores, roadside stands, farmers' markets (both seasonal and all year), and other beekeepers so you know what you are up against. The National Honey Board Honey Locator is a great resource for this (see page 25).

You probably know what bad honey tastes like, because you've made some in your day. It could have a burned, off flavor from something out there, it could be fermented, or it could be a blend of something evil and something good that has a wonderful first burst but an aftertaste that makes you gag. Or it could be so bland that sugar syrup is more interesting. You don't want to be selling that to what will be a lost customer. Always sample before you stock. Make sure the flavor works, and don't worry too much about the color.

What honeys to sell? First, look at what others aren't selling and then ask why. Is there no demand or no supply? Look at all the things sold by other operations on the National Honey Board site; everything imaginable, and a lot you hadn't thought of, is for sale somewhere. So find out why local buyers or local sellers don't buy chunk comb honey, orange honey, honey sticks, or whatever it is. If it's a legitimate reason, you won't be burdened with having inventory you can't sell. But if it's a lack of experience on the part of your customers and, thus, low demand, you can change that. If it's a lack of supply, you can change that, too.

Once established, you should probably have a diverse inventory, in both varietals and types.

All of this works for both in-store selling and selling to those customers you deliver to—schools, stores, bakeries, and the like. Deliveries cost time and money, and wholesale sales reduce the margin on every jar sold. So your business can be all store, a mix of store and delivery, or all delivery. Your ultimate mix should be profitable and repeatable.

Obtaining Honey

If you're keeping enough bees to sell all the honey you want to sell, you can skip this part of this chapter. However, if honey is part and parcel of what you do, then you'll eventually be buying product from other beekeepers, packers, brokers, or somebody, at least in some years. You have customers who sell your honey, and if you don't have honey—due to crop failure, bee failure, floods, or the like—they will have to find someone else. Then when you do have honey again, you will have fewer customers. When you weren't living off your honey sales, a dropped case, a customer who vanished without paying a past-due bill, or a winter loss of 90% was something you might have been able to absorb. After all, you had a day job. Now, every jar of honey has to make a profit, not just pay for itself. Every jar has to cover its production cost, operating costs, the insurance needed to cover anything that could go wrong, debt and its interest, and of course your labor.

When you are buying honey from other beekeepers, you need to make sure it's as good as what you normally make. What's in the bottle or pail you sell is your reputation, your good word, your honor, your business. Your customers expect it to be your quality, and that's all there is to it. So knowing what you are buying is critical. When you know the beekeeper and trust the product, your liability is probably close to zero but when you don't, things change. How far do you go and how much do you spend to make sure you know what's in those pails? Because this is your reputation, testing procedures must be part of your purchasing budget. There are different levels of analysis, from simply tasting and looking at samples for color, to measuring for moisture, to pollen analysis, to high-tech adulteration sampling. Of course, the age-old grading system still has some value in determining quality, too. You can state, for instance, that you will only purchase honey that qualifies as USDA Grade A. These grades, even though incomplete and inadequate, can allow you to easily pass on some purchases without spending money on analysis.

How far do you go to make sure you know what's in those pails? Because this is your reputation, testing procedures must be part of your purchasing budget.

TIP: Here are a couple more pieces of advice about purchasing barrels: Even if you know the seller, have dealt with them before, and there's a level of trust, you still need to be careful of all the things that can go wrong—things you can't see with the lid on. Moisture and flavor issues are at the top of the list. Is the honey in the barrel the same as the honey in the sample that was sent to you? Make sure it is before you close the transaction.

GRADES OF HONEY

Following is the USDA summary of the grading system that has been used for years. There are flaws, but for most of us it still works in sorting out the good from the bad.

Grade A: Flavor and aroma is good. It has a good, normal flavor and aroma for the predominant floral source or, when blended, a good flavor for the blend of floral sources and the honey is free from caramelized flavor or objectionable flavor caused by fermentation, smoke, chemicals, or other causes with the exception of the predominant floral source. Clarity is essentially clear, but it may contain air bubbles that do not materially affect the appearance of the product and may contain a trace of pollen grains or other finely divided particles of suspended material that do not affect the appearance of the product.

Grade B: Flavor and aroma is fairly good. It has a fairly good, normal flavor and aroma for the predominant floral source or, when blended, a fairly good flavor for the blend of floral sources and the honey is reasonably free from caramelized flavor and is free from objectionable flavor caused by fermentation, smoke, chemicals, or other causes with the exception of the predominant floral source. Clarity is fairly clear, but it may contain air bubbles, pollen grains, or other finely divided particles of suspended material that do not seriously affect the appearance of the product.

Don't even bother with Grade C. It's not worth the time it takes to see how bad it really is. But there are other aspects that you should be aware of, and use if necessary, to sort the good from the bad. Flavor and aroma can be somewhat subjective, but if you are buying, it's you who decides. And, there is some level of legal definition used when grading, also from the USDA:

6.5 Flavor and aroma. The honey flavor and aroma will be appropriate for the source specified by the user. The honey must be free from objectionable or foreign flavors or odors, such as fermented, yeasty, moldy, effervescent, bitter, etc.

All of these standards are used on labels, and the reason is to protect the consumer, which, incidentally, includes you. When purchasing, country of origin shouldn't even enter the picture, but if it does, here's the basics of how and when you need to use it.

6.2.2 U.S. Grades and Country of Origin. When honey labeling is marked as a U.S. Grade, the labeling must also include the countries of origin on the label. In accordance with the 2008 Farm Bill—amended the Agricultural Marketing Act of 1946 to require that packaged honey bearing a grade mark or grade statement, continuous inspection mark or statement, sampling mark or statement, or any combination of marks or statements of the U.S. Department of Agriculture, must also bear the countries of origin of the container of honey. This listing of the countries of origin must be printed legibly and permanently, placed in close proximity, and of comparable size to the certificate, mark, or statement, and preceded by the words "Product of" or other words of similar meaning.

ADULTERATED HONEY

And then there's plain old everyday fraud. Some sellers will add sugar or corn or rice syrup, or even water, all of which are detectable, but at what cost? Adulteration is the most common fraud committed in this industry, reducing the cost to the seller and certainly the quality of the end product. The cost to detect it is pretty high, not only in dollars, but in the damage to your business if you miss it and it is detected by a customer. Moreover, your business can even take a hit if it's found that your supplier cheated someone else, and you are guilty by default—yes, even if you had that last batch you got from him tested and it was found clean. The fact that you dealt with a criminal will taint your once-perfect record. Don't buy from people you don't know or don't trust. Lost sales because you don't have product are not nearly as expensive as lost customers who no longer trust you and your honey.

According to the USDA:

Adulteration of honey is the addition of other sugars, syrups, or compounds into honey to change its flavor or viscosity, make it cheaper to produce, or increase the fructose content to stave off crystallization. According to the Codex Alimentarius of the United Nations, any product labeled as honey or pure honey must be a wholly natural product, although different nations have their own laws concerning labeling. Adulteration of honey is sometimes used as a method of deception when buyers are led to believe that the honey is pure. The most common adulteration-ingredient is clear, almost-flavorless corn syrup, which, when mixed with honey, is often very difficult to distinguish from unadulterated honey. Rice syrup is also commonly used and is detected in a similar manner.

Isotope ratio mass spectrometry can be used to detect addition of corn syrup, rice syrup, and cane sugar by the carbon isotopic signature. Addition of sugars originating from corn, rice, or sugar cane (C4 plants, unlike the plants used by bees, and also sugar beets, which are predominantly C3 plants) skews the isotopic ratio of sugars present in honey, but does not influence the isotopic ratio of proteins. In an unadulterated honey, the carbon isotopic ratios of sugars and proteins should match. Adulteration levels as low as 7% can be detected, but that 7% can be the profit margin on your purchase. Still, it's the best test there is.

7.1 Analytical, physical testing, and reporting requirements. When specified by the user, the following analytical and physical requirements for the honey must conform to those listed by USDA requirements. The analyses must be made in accordance with the appropriate methods from the AOAC International Official Methods of Analysis (OMA). Any result not conforming to the analytical requirements may be cause for rejection of the lot.

6.4 Additional ingredients. The honey will not contain any additional food ingredients including food additives.

BLENDED HONEY

There's one more type of fraud that you have to watch for: blending.

Say you have a barrel of soybean honey, which tends toward having very little flavor. Because it's only sweet, you can add a pound of clover honey and suddenly you have a barrel of honey that tastes like clover, with a higher asking price and better flavor profile than soybean. It's all still honey, nothing illegal has been added, and if the seller simply says honey and not clover honey, it's not fraud. Kind of. And, to you, it seems like clover, smells like clover, looks like clover, tastes like clover . . . and you've been cheated.

Blending can benefit you if you have customers who want their honey to taste just like the honey they bought last time and the time before. That's commodity honey. It's a legitimate way to do business, as long as you are honest with both yourself and your customers. If they are more than satisfied with this and you've given it a generic name such as Autumn Blend (which would probably be a darker, tad more flavorful blend) or Summer Breeze (lighter, milder than autumn), then you are making no claims as to origin or variety. But they will be able to find it again because of the name you've given it. So there's blending, and there's blending. And they aren't the same. Be careful when you buy, and be honest when you sell.

Selling More Than Honey

We aren't going to spend much time dealing with actual honey production, but we should include some mention of keeping bees alive because it is the most expensive part of your bee budget. Right off, and depending on your agenda, labor is probably the biggest part of that expense. How and what you do, relative to making honey, rests with what else you do with bees, and if what else you do is mostly what you do, then paying attention to these details makes sense. If you pollinate early, mid, or late-season crops, your mite treatments will differ according to which part of the season you have bees in the orchards. If all three, you don't have to worry much about honey because you will, in all likelihood, be a lot more worried about keeping stress and mites to a minimum, and feeding will be right at the top of your list of things to do (are you making enough pollinating, in time, to keep buying all that feed, by the way?). You may be able to capture a honey crop on one of the drops, or at a holding yard between one of those crops, so timing for honey supers (after a treatment and before the next move) will be critical. If you intend to make more bees with spring and summer splits, pollinate and make honey, your mite scheme will be busy, but doable. Walkaway splits with a queenless period will help with mites, but may or may not influence nectar collection and, thus, any other chemical treatments you were considering. And every season, this mix will change due to weather, bees, farmers, beekeepers, and luck.

By now, you know how to produce honey. You know how many more colonies you will need to accommodate planned business increases. You know how much acquiring all that will cost and the kind of time it will take to get there. For example, just the

Are you going to let this very expensive piece of equipment sit unused for fifty weeks of the year?

boxes: Will you buy existing? Buy new? You can add up the assembly time (will you buy assembled and take no time, or buy knocked down and take lots of time but less-up front money?), the time it will take to find new beeyards, the time it will take to haul hive stands, and establish time. Splits have a schedule for build time, packages have a different schedule for build time, there's feeding time no matter what, you may need extra harvest time (maybe this year, maybe not), and you may need extra winter prep time.

Your equipment and labor costs are adding up. Do you have that saved up already? You have to have it before you get all that extra honey to pay for it. Borrow on credit and pay the interest for a season; there's a hidden cost. How are you going to be able to buy what you need to expand? Your very good business plan will have that figured out (see page 13).

You can get by with what you have, I'll bet, at least for a season or two, so you have more honey that second season to absorb the cost of new equipment. And by this time, you'll be looking at potential new equipment in a somewhat different light. It will be less a question of if it's good enough, and more a question of how much better you can make it with better equipment. How many problems will that new gadget solve, and how much time and labor will it save? In the light of day, the gadget may pay for itself in weeks just in the labor it saves you. Don't write off your time.

Now, consider that the equipment you are using is active for only a short time each year. The rest of the time it sits unused not earning you money, but still needing to be paid for. Can you remedy that? Yes. Provide services. You can likely even bring this into the operation with your existing equipment because you probably don't have that nine-to-five job anymore, so you should have the time to do it. (But, see about honey house regulations on page 37. You may have the equipment, but not the necessary permits.)

You can essentially become a packer for other beekeepers who do not have the efficient machines or the extra time that you now do. They bring the honey, the bottles, and labels, and you fill and label for them. Yes, they may be competitors, or at least on your radar. Nevertheless, they will be paying you to do their work and paying your bills. Choosing these people specifically because they don't interfere with your customers is a good move. Distance and type of product are good considerations, too, but you can make it work. Payment can be by the case, by the hour, or by the barrel or pail emptied. You know how much an hour your operation has to make, every hour, every day, and you can calculate from that figure. Another benefit of your business plan.

Another income generator you might consider is custom extracting. You use your equipment and their supers full of honey to put into their pails. This is especially nice for smaller operations that don't have the equipment you do, or for the hobbyists who don't have the time to devote. There are some precautions to consider. Primary among these is contamination of your equipment with foulbrood. Because you'll get only honey supers, you don't have brood frames to inspect, so you won't know. You can insist on something from a local inspector indicating clean equipment, or simply clean everything yourself when done. Slapping an antibiotic on your colonies isn't in the equation anymore. And what do you do with uncapped frames? Leave them? Extract all together and return as feed to the beekeeper? Make sure you have that figured out. And what about crystalized honey in the comb? It stays in the comb.

Don't forget to check local food-processing regulations to make sure you have all your i's dotted before you start. Some places won't care, and others will close you down if you are processing food without the proper permits and inspections.

Payment can be cash per super extracted, some portion of the honey if it is good enough, some portion of the cappings wax, or simply cash per hour it takes. Some customers take longer than others, especially if you have to spend time vacuuming bees off windows and lights.

Private labeling is also an option when you have more honey than you can sell to your customers, but you don't want to simply put it in a barrel and dump it on a packer's loading dock for low-ball prices. With some precautions, you can put it in jars for other operations—your jars or theirs, their labels and cases. Because the honey is top shelf, they may take some of your customers away, if you let them. Don't. Keep them separate by distance, quality of markets (big box stores vs. upscale markets), label (their Ohio Wildflower vs. your Tulip Poplar), or the like. And remember, time is money, so charge for that, too. Although, if they were already a competitor, you wouldn't be talking to them in the first place, except at the monthly beekeeper's meeting.

Certainly, there are precautions that need to be taken before bottling your honey for another business. Who is responsible for problems in filling weight? What about crystallization on the store shelf? Who delivers? What about the timing of the job? This can quickly become as complicated as a pollination contract. And here, *contract* is the key word. Spell out everything that can go wrong and what should happen when it does in your contract.

The person you are bottling for is putting his reputation in your hands and you better well do everything you can to protect it. Conversely, you have little exposure if there is an issue unless it is the quality of the honey (flavor, moisture, color) that comes into question, and then the contract will have spelled out what you should have done. Don't get me wrong, this isn't a snake pit you can fall into, but do take precautions so neither your customer nor you have a problem down the road.

SELLING HONEY PROCESSING EQUIPMENT

What about the honey processing gear you will carry? The smallest items—uncapping knives, brushes, escapes, fumeboards, and the like—are easy to keep in inventory and should be on hand because they will be an "I need it now" item when somebody rushes in on a Saturday morning. Even expensive items should be available, at least in season. This includes pail heaters and pails, valves, filters, and a couple of small extractors. But how many $1,000 extractors can you keep out back, or what about those $4,000 uncapping/extracting/wax-handling outfits you see in other catalogs? Do you even want to deal with this level of business? You can have these available from the manufacturer or operation that you are a dealer for. Mostly, as you are probably aware, these are specialty items made by dedicated manufacturers, and your customers will buy the bigger, more expensive items directly from them and have them shipped. And that's not a bad way to handle this.

CONTAINERS AND LABELS

Next, you must have the containers your customers' honey will go in. For those that make enough honey that they can consider buying a pallet's worth of glass, you can still compete with a distributor because you should have a much greater variety of containers. The standard queenline jar, whether glass or plastic, is still what many use, but fortunately many beekeepers are branching out with other shapes and sizes. And you can introduce them to even more. Basic customers may stick to the queenline, but there are now cylinders, embossed cylinders, hex shapes, skep shapes, glass and plastic bears, mason-like bottles, muth jars, and many smaller jars used as wedding or party favors. Plus, there are many more sizes available now in both glass and plastic, so customers can price them differently. Same price, smaller bear works in many markets.

There are as many containers as there are beekeepers—and labels to match.

With these are embossed and painted lids, corks, gift packs, comb boxes, creamed honey tubs, and more labels that most of us can imagine to fit all these new containers. Custom labels will run between $0.10 and $0.15 each, depending on quantity. The lid, label, and jar should total about a $1.25 or so,

keeping it about 12–15 percent of their retail price. Fixed up with your special containers, your customers will all be unique. The days of pallet loads aren't over, but better ways of presenting honey are here to stay.

Your Labels and Containers

Let's take a long look at the label, and while we are at it, the containers you use for your products. A good bet is that the label you are using is one of the basics from a bee supply company and you've been using it forever. Another good bet is that you'd rather cut off your right arm before changing that label, because absolutely everybody in the world knows your brand and your honey and you because of it. Right?

Take a long look at your label. Why is it special? Is it special? Now, go online and search for "honey label design" and take a look at the thousands of different labels all doing the same thing your label does. Basically, all the same information is on every label: weight in grams, pounds, and ounces; owner, nutrition information, contents, country of origin, a warning about kids under 1 year of age, variety, and all the other legal terminology.

Some have top and back labels, some have wrap-around labels and top labels, and some have neckties. Some use the same label on every size jar; some use the same label, but bigger or smaller. Some have boxes, white lids, black lids, gold metallic lids, plastic jars, yellow plastic lids, black or white or red spouts, or flip tops. There are quart or pint jars, glass jars, plastic jars, hex glass jars, round jars, square jars, little tiny jars for wedding receptions, plastic bears in small, average, or large, safety seals from lid to jar . . . . There are, essentially, infinite combinations of containers and labels. If you're still using glass queenline jars with white lids and those 12-ounce (0.35 L) plastic bears with flip-top openings, you either have no imagination or your customers don't like change.

Both your container and your label should reflect the market they are being presented in. Quart jars with dented lids and bee supply one-size-fits-all labels in a Whole Earth Foods store fit just about as well as a 12-ounce (0.35 L) hex jar with a gold cap and front, back, top, and safety labels does in a small-town farmers' market next to some summer squash. Have some respect for your customers and your product.

Take time to re-examine your containers, lids, and labels to match them up with the market you are selling to. You may, in fact, need several varieties.

While you are looking at those thousands of labels, take a good, long look at the containers they are on. The people buying your honey at the boutique grocery store aren't the same people buying it at the mom-and-pop corner grocery on the edge of town, are they? Why treat them the same, and why charge them the same? Why look overdone to one group and like trash to the other? Label design is a cost, printing is a cost, and bottles that cost more than a plastic queenline are a cost, but the price you are selling these items for accommodates that cost. Know every outlet's customers. You are not a one-shape-fits-all business.

Internet Retail: A Hybrid

I'm not sure if selling honey on the internet is retail or a sale by a faceless soul behind the keyboard. You do have to be registered for sale in other states, but the inconvenience is worth it to provide an online opportunity for your customers. Even if you never meet or see or talk to an online customer, there is some aspect of retail on a webpage. It has to do with letting people get to know you. If I'm buying your honey, and you say it's local, raw, and blueberry, it's only so many words under the picture on the screen. But if you've told me your story—about you, your family, your history, your honey, and your pollination business, with photos and videos of all of this, beeyards, kids, a honey house tour, all your products, and your blueberry farmer's fields—now I know you and your operation. I'll have a feel for who you are, what you do, how you do it, and why.

Include praise from previous customers. Make sure there's a place where customers can make comments (and that you have control over it so you can remove trash), and then use them as much as you can. Word-of-mouth advertising is free, and it's the best there is. Being able to tell your story and letting customers respond is about as close to retail you can get on the internet. Start thinking and planning now for that retail show on the web. Really, make it a show. It doesn't have to be slick and professional; in fact, a tad folksy is even better.

 An additional method to consider here is to link to Facebook Live videos you have posted. These can show field days, harvesting adventures, package day crowds, you working your bees, or anything that gives your business a personality. Keep adding new live videos—some every week if you can— at least seasonally to show the growth of your business, highlight changes in your store, or summarize an event that just occurred.

Selling Honey

Whether it's by pail or pallet, selling honey is likely going to be a large part of your business. When it comes to selling honey, it's not so much what you are selling (though that is half the battle), but where you are selling. And importantly, where aren't you selling that you should be?

SELLING BULK TO OTHER BEEKEEPERS

Businesses that buy and resell honey from and to other beekeepers play a special role in the honey business. They are in a position to know producers who don't like the art of selling very much and sellers who don't have enough bees to produce all they can sell. These in-between players make connections and keep the flow going. The business isn't so much what you know, but who you know. Generally the margin made on a pound isn't great, but the investment cost is also very little. If anything, their cost is in delivery, and in having money tied up waiting to get paid. If these in-between sellers are solid, they can wait to get paid until the buyer finally sells what was delivered. Most don't recommend this because of the risk of not getting paid at all and not being able to recover the product. The other risk is inventory cost when buyers have more than they need and there's no demand for product sitting out back in a warehouse. But it starts with who you know, right?

If you go this way, some of the wholesale customers you will be working with include users who need timely deliveries, the right-sized containers, and usually consistent color and flavor. These include mead makers, breweries, wine makers, food service, bakeries, and other types of food manufacturers. Some will be easy to work with because, like you, they both buy and sell mostly locally.

The easy part of this is the containers you will need: pails, maybe gallon jugs, barrels, or even totes. And no labels—or no fancy labels, anyway. The hardest part of this is always having exactly the right honey ready at the right time. If your customer has been using a clover/sunflower blend for the last century, and you and every beekeeper in the Midwest had a sunflower bust this year, you'd better find some clover/sunflower blend fast, or they will find it elsewhere. They have a tighter schedule than you do, and their customers are just as fickle as your customers are.

To accommodate this, you need to think a year ahead for sourcing or producing that blend. And if this customer is the only customer you have . . . need I say more?

Other buyers in this category may not be so picky, but know in advance if they are. Deliveries may be an issue, with just-in-time pails being part of the deal.

Other beekeepers can be a lucrative market, too, if you can produce enough of the right kinds of honeys for them. In fact, you may be a perfect source for them because you are local, trustworthy, and don't sell in barrels. Certainly don't discount this opportunity to move some of your honey on a regular basis, especially the more common varieties you have. This isn't the cheapest wholesale you can make because your honey has to be ready to bottle the way the customer wants it, whether that's completely raw or crystal clear. It's their market and their product when it goes in the jar. Don't argue.

TYPICAL WHOLESALE

Because of time, location, or inclination, you may choose to never sell a jar of honey directly to a retail customer. Instead, you can deliver cases or pallets to a warehouse, store, or manufacturer every week or so, keep it strictly wholesale, and let them handle the customer service. If that's what you've been doing, it works, your budget balances with the margin you are making, and this is your comfort level with people, you have a good start. Look at the places where you are now selling some products, where you already have your foot in the door and they know and trust you. Your honey will sell in these locations because of the consistent quality, not price. Why not use these relationships to expand what you sell in each location? If you are already delivering to a store once a week or so, there's no extra delivery time or cost. It

Pails are an easy way to sell a lot of honey.

will be more preparation perhaps, but you are probably already selling different things to different people, so you already have these additional products ready. Consider the following:

- Could you sell more of the same local products you are selling now, say three rows of 1 pound (0.5 kg) jars instead of one or two, reducing your number of trips to the store, the time you spend delivering, and missed sales when your products are sold out?

- If you are selling glass, introduce plastic with a spout and reflect the cost savings in the price, so there is incentive for those few who do look at price.

- Could you sell the same local products in different sizes? If 1-pound (0.5 kg) jars are selling like hotcakes at a retail price of $8.00, your production cost (honey, overhead, etc.) is $4.50, and you need to make a profit of about $1.50, your wholesale cost is $6.00. The store seller adds his overhead costs of $2.00 plus a reasonable profit of $2.00, for a retail price of about $10.00 or so.

For a 2-pounder (1 kg), maybe your cost is $9.50 or $10.00, plus added profit of at least $3.00 for a wholesale cost of $12.50 to $13.00 or so. The retailer would mark it up another $3.00 for overhead and profit for a retail price of $15.00 or so. Both you and the store owner have to have every jar sold pay for itself, overhead, and profit. The same math works for smaller containers.

- What about using the same size jars, but introducing different products? For example, it could be your local honey but different seasonal varieties, say spring, summer, and autumn, all with your name and label so buyers already know the quality.

- Have you tried what may be considered exotic products, such as creamed honey, honey infused with peppers or other flavors, or comb or cut comb?

- Certainly have a raw, and usually crystalized, product along with a barely strained but not crystalized jar, same label and color. Maybe price the raw differently.

- What about selling beekeeping products, such as candles, blocks of wax, and jars of pollen?

- One way to differentiate products and, thus, make more varieties available, is to give them a purpose. Best BBQ Honey in Ohio might work. Terrific for Tea, Baker's Best, and Sweet Home Breakfast all tell customers the best way to use your honey.

- Could you have a brochure available that tells about your business history, the other work you do with bees, how your product got its name, what you have for sale, where you're located, your hours, and all the rest of your promotional material?

- Could you add a top or back label with some of this or other additional information? For example, you could include a short definition of what "raw" means, granulation information, or three suggestions for how to use honey instead of sugar for baking.

Some wholesalers might not be reselling your honey but, rather, using it to make something they are selling. These can be ideal customers because they are usually consistent, they're businesses so payment is usually timely, and working with them in dual promotions is productive. Seeking them out is not a bad idea.

STRICTLY RETAIL

You may have been doing this successfully for some time now. Perhaps you see good markup and enjoy the process of talking and selling and helping people with what they think they want. Farmers' markets, for instance, both seasonal and year round, offer this opportunity. You can have direct, face-to-face, you-want-honey-and-I-have-the-best encounters all day long. Get up early, set up, sell, sell, sell, take down, clean up, and go home.

But this can be tough, simply because of the time involved. You can only be in so many places at a time. All-day selling means all day not keeping bees, harvesting, delivering, cleaning, building, pollinating, and whatever else you could be doing to make more money. Granted, there are celebrations and events at which you make more in a day than you used to make a week working 9 to 5, but how many of those can you have a week or a summer?

So, spread out. Be in more places at the same time. It's not hard when you think about it. First, find several good farmers' markets in your area; some may be distant, but keep it close enough that you can say the honey is local. Even if there is another beekeeper selling, your honey and that honey won't be the same. But how can you be in two places at the same time? In a word commission.

A good salesperson can sell ice cubes to penguins, so selling your quality honey should be simple—if you make it worth their time, of course. That salesperson has, most likely, lots of time to prepare, because the person you are looking for is looking for a seasonal one- or two-day-a-week job. Make it worth their while even more by using a sliding commission; the more they sell, the higher the percentage they get on every bottle. That's an incentive program that can

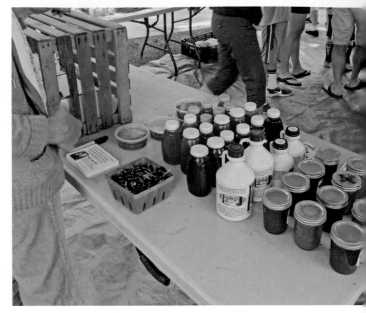

Working with CSA farmers, or other farmers' markets, you *can* be in two places at once.

work well for you, as long as you keep the profit you need in mind.

What might be easier is essentially outsourcing your sales force. Work with vegetable, produce, and bakery vendors in farmers' markets in nearby towns. Have them sell your honey right along with their products. This gives them additional attractive products to sell, and keeps the products they are selling essentially local (watch the regulations; it may not work). You could even private-label it for them, putting the name of their company on the label, but still saying "Produced in Ohio." You can sell them the honey directly wholesale or work on the commission basis. See, you can be in two or more places at once.

One trick for a farmers' market booth: Make sure there are enough people. You need a seller, a filler, and a cashier. The filler and cashier may be the same person, depending on volume of customers and difficulty of filling, but you absolutely can't be selling your honey, making change, and filling bags or shelves at the same time.

Over time, your brand will be recognized, and in demand, in more places than just hometown stores, and you'll be able to open additional wholesale outlets. Customers get to know you during the summer, but want your honey all year, and this creates additional demand during the off season. Getting your products out of your backyard is a win for you, the people you get to help, and the customers who now have access to a quality product they didn't know existed.

PERSONALIZED WHOLESALE SALES

If wholesaling to local stores and other outlets is the majority of your sales, you can personalize these outlets by doing tastings and displays and offering teaching moments in their stores. You still sell them the honey at your wholesale price, but you offer to "help out" a little. You can hire or commission someone to do this in other stores for you, too.

You need a small table, three or four of your honeys including some comb if you have it, tasting spoons, maybe some cheese and crackers, and lots of in-store and newspaper promotion in the preceding weeks to let folks know when the beekeepers will be in the store. Bring a helper, if at all possible. You can handle four times the people if you have someone cutting, dipping, pouring, and cleaning; plus, they can help hustle people over to your table if it slows down.

If you can, get the owner to offer a sale price for the day, and don't be stingy with the free samples. Let everybody taste everything, twice if they want. Have lots of pairing samples with you. Bring asiago, stilton, cheddar, and gorgonzola cheeses so you have mild to strong cheeses to go with strong to mild honeys.

Bring more than cheese, too. Try dark chocolate for strong, dark honeys, melon for mild honeys, pineapple, cornbread, all manner of different crackers featuring herbal flavors, and any and every kind of fruit in season. This works especially well if you are in a grocery store where the taster can head over to the produce section and get more to go with the honey they just picked up.

If you have comb honey, have small cut and drained samples ready with toothpicks or crackers to put a chunk on. But here's the world's worst-kept secret: the absolute best way to sell comb honey is to dip a chunk in a pot of hot chocolate ice cream topping. Just a dip. Freeze the comb chunks the night before and keep them in a cooler during the day. And make tasting them kind of secretive. If somebody asks what "that" is and points to your comb honey, give them a sample and see what they think. Then look around and kind of whisper, "you want to try something really special?" When they say yes, get out the chocolate, look around again and do the dip. Cover just the bottom of the chunk or you'll overpower the honey taste. But the two combined . . . you won't have enough comb honey to sell.

Just like with farmers' markets, you can organize commissioned salespeople for this, based on increased sales to the owner, so you can be in more places than you thought possible and you won't ever have to leave home.

RETAIL AT YOUR STORE

Having honey for sale in your store can take all manner of forms, and you need to carefully watch inventory costs as you explore what this market can do for you—and to you.

The easy route is to sell what you sell to others: the same varieties, same containers, similar prices, and same labels. No imagination, no extra work. The harder, but more lucrative route is to sell what you sell in other locations, but also have different containers, labels, varieties, prices, kinds (comb, creamed, chunk, exotics, colors, and the like), and products (beeswax products, pollen, propolis products, candles, and the like). What customers see on the store shelf or farmers' market table is only a sample of the goodies that await on your shelves. But they have to know that. How will you tell them? Top label or necktie? Brochure in the bag at farmers' market? Actual ads on social media, in your catalogue, or in the local paper? All of the above?

For in-store demos, you can use frozen chunks. For show, keep some comb on a rack over a drip tray and cut chunks to dip in chocolate sauce.

The customer you are trying to attract, really, isn't the beekeeper who stopped in needing a mite treatment chemical and had your display catch his eye. Your potential customer actually came out because of the brochure, top label, or newspaper ad. That's who you want. Of course, you have what they remember from the market, but suddenly there's a display of more than they've ever seen, heard of, or tried. Maybe it's four or five different products, or maybe it's a couple dozen.

Then, right next to that display, you have the rest of your bee products. These won't interest many beekeepers, who make them at home, but they will interest anyone who wants an "all-natural" cosmetic product, a long-burning, clean-burning beeswax candle, or a propolis tincture you can't get anywhere but on the web for that hot-handle stove burn. Profit margin can get really generous on this shelf.

These products can be made in-house, because you already have the wax, but if you don't have the time and talent to make them, this is the perfect place to sell on commission or simply to buy and resell. Either way, the margin should be pretty favorable to you.

This is the time to consider want versus need. Customers in a farm-supply store *need* things. Customers in a Macy's *want* things. There's a big difference. Most of your customers will come to your store needing things: smokers, honey supers, queens, etc. Your retail honey and bee products display needs to make them want something: a jar of exotic honey, for example, or that perfect set of candles.

Put up clear easy-to-read signage in your store. Your signs should give the customer all the information they need to make the sale: What is this stuff? How do I use it? Why should I buy it? What does it cost? What is it made of? Where did it come from? And make sure they know this is the only place you can get these products.

The customers who saw you last summer will come because they want something. They want more of that honey they bought before, but when they see what more you have, the want button has already been pushed and you've got them. Use that shopping mood to your best advantage. I want that honey, but wow, facial cream with beeswax, candles for supper, and comb honey for my parents! That's what you want for those customers.

PAILS, PACKERS, AND INDUSTRIAL SALES

We've talked about selling pails to other beekeepers; your honey is a lot like theirs, and they can use it without fear of being cheated, and without offending or surprising customers who are used to a familiar product.

The days of shyster packers are pretty much over but new ones keep showing up. "Buy low and sell low" is the motto and, for some, the quality of your product is second to the price. How you would buy honey if you had to? What do you look for when you need more and all of yours is gone? If your packer is as fussy as you are, you can be comfortable with

This much honey puts you in a different league than most beekeepers.

This works two ways, of course. You have an obligation to deliver what you say you will. It starts with your storage. How long has that honey been sitting there, and how hot did it get? That's probably the only thing that can go wrong, other than the lids not being snug. But long periods of too much heat will change what you had into something your customers don't want. And the transition from not having enough to having too much will be variable the first few seasons. If you have a very good crop and not enough customers, what you have may have to sit . . . in an uninsulated garage and cook?

"Send a sample," says the invitation from the packer. So what do you send? Make sure it's representative of what's in the rest of your pails, in terms of color, flavor, and moisture. Be as fussy when selling as you are when buying. That way, you won't end up having to take it all back, or take a price drop because it wasn't as good as the sample(s) you sent.

Consider the differences between your business and the business you're selling to. Their customers tend toward wanting a consistent product—the same color and flavor every time—so blending is their practiced art, and your honey is simply one of many that produces that consistent product. Extreme heating and filtering to remove any speck is necessary, because when that crystallization happens, grocery stores won't put up with it. They can't have a bottle looking spoiled or a customer bringing back a bottle that's gone bad.

These folks are as fussy about their finished product as you are, but it's a different finished product. Maybe it's not as good as it could be, but it's a saleable product, and consistency is essential.

them. You'd look for clean containers, honey that isn't heavily filtered, a delivery that's accurate and on time, payment on schedule (that can be a tricky item though, so be careful), honest assessment of variety in the barrel, and no damage in the slightest (nothing burned, stored too hot, or with any off flavor for any reason, in any of the containers). That's what you should be looking for, and if the person buying yours isn't that fussy, what else aren't they fussy about? Follow that gut feeling.

Beeswax Products

Beeswax is a given in a beekeeping business. Your bees will make some for you if you have bees. Your beekeeping customers will have wax to use and to sell, and they may want to buy more. Buying and reselling is an easy part of the job. There's little markup, but, depending on how fussy you are, there's essentially no investment. And it's being fussy that counts.

A good way to buy bulk wax from your customers is to require that it meet minimum cleanliness standards. It should be melted, filtered so there is no debris, and set in common, easily obtained molds to cool. The pans commercial outfits use to pan out wax are ideal. They are standard in the industry, and they're easy to use. You can sell them to your customers, and their wax and everybody else's wax will come in standard sizes and shapes. What you'll get is clean and uniform.

What about color? Will you pay more and, thus, charge more, for lighter wax? Yes, you will. Here's one way to do that efficiently: Look at a wide variety of wax samples, from the lightest to the darkest you can find. Find a sample that's about two thirds of the way to the lightest, and one that's two thirds of the way to the darkest. Those are your benchmarks. Compare any wax that comes to your benchmarks.

It starts here.

It will be lighter than the first, darker than the second, or medium—that is, in between your two benchmark colors. Those are your three colors, light, medium, and dark, and you'll pay appropriately. It's less important how you do this than how consistent you are. Customers remember. Make a photo of your benchmark color samples, post it prominently, and there will never be an argument.

What about all the wax products you can make? Simply, how much time do you have? Test the waters a lot before you spend significant time or energy on

Bulk wax blocks are easy and fast but offer no added value.

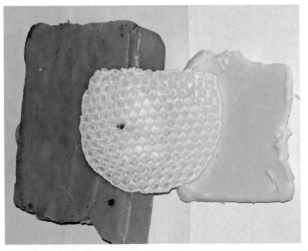

Waxes of different qualities.

Beeswax: valuable and variable.

seasonal or farmers' market exposure. This goes out the door if someone in the family gets up every day thinking candles and lotions, however. Then, you focus on these products, sell them in other outlets, and even to beekeepers to resell. Don't waste talent, but know the cost.

There are a couple of ways to capitalize on a soft market or not having enough time. First, sell other people's products in your store on consignment. This works for every beeswax product made, including candles, figurines, lotions, and potions. Let someone with time and skill make these for you, so you can simply display and sell with no risk, while it provides your outlet with additional attractions. This works both ways. Make contacts with other craft outlets that sell on consignment or, better, are looking for exactly the products you are making and will buy them from you. There will be some time and cost to make these, but you won't need shelf space. And they may do quite well in a different location, customer base, or price market. Beeswax is a part of beekeeping. Use it to its fullest potential for profit for your business, but be careful about your exposure.

Then there's the lotions and potions market. It tends to be a lot like the candle market in terms of seasonal sales and demand. Selling on commission is one way to handle it. Private labeling is another; that is, somebody makes it for you with your name on the label. Either is a good way to have everything and

these products. If you have a fair stop-by or walk-in nonbeekeeper honey trade already, they might do well because your customers are not looking for supers and extractors. If you're strictly a beekeeper's business, though, they probably won't. The mark-up on candles and figurines is pretty good, considering what you pay for wax, but sales tend toward

Candles.

Lip balm.

Christmas tree ornaments.

anything bees in your outlet. Soap falls into this category, too. Make your own if you have time and talent, sell someone else's, or have them make it for you.

Don't ignore the opportunity to sell supplies for these, however. The oils, waxes, molds, wicks, and containers have to come from somewhere for your customers who make these things. Being able to buy one or two is much better than a dozen or more that don't ever seem to get used.

You can also sell how-to lessons on making candles, lotions, potions, soap, and ornaments. You charge for the class, and it includes enough materials for each student to make whatever it is you are teaching them to make that evening. You, someone in your family, or the person selling on commission or private labeling teaches the course. Your location, already known to all the potential students, offers refreshments and help out to the car, afterwards, because your students will be carrying what they made in addition to everything they bought while they were there.

There are dozens of beeswax products to make, and teaching folks how is a no-brainer. Offer classes for each product or each class of products, every few months so people can keep coming back to learn something else, or to learn a more advanced technique on the same product. Every year or so, start the process all over. After a short time, you'll have several levels on several products going simultaneously. The objective, of course, is have your students/customers return, pay for another class, buy the other things you have, and become a preferred customer; after all, the classes are held when you're not normally open, all the appropriate products are on special that evening only, and you do make the best honey goodies on the planet they can't get anywhere else, unless they want that cookbook you have over by the rest of your books (which will show them additional ways and techniques).

Teach a class with a live demo.

Beeswax soap. Ideally you'll have a large selection.

Your soap labels need to be both attractive and functional.

Another use for beeswax, and an opportunity for another class, is to make Ukrainian Easter eggs.

You can sell beeswax soap, and you can offer classes in how to make it.

The cost of this is nominal, and you more than make up for it with the modest tuition, which covers the cost of the instructor's time, the raw materials, the goodies, and coffee. And, even though you make less margin on the items you put on special, you wouldn't have sold any of them unless you had the class. I'll repeat: it's a no-brainer.

This opens another door. Because you have all these handmade, one-of-a-kind products, why not specialize in corporate gift baskets? They could feature all manner of beeswax products, several kinds of honey, and perhaps some baked goods. Because of your classes, the word is out about the quality of what you make. These aren't inventory items. Know the cost of the components, the time it takes to make them, and the time it takes to assemble, decorate, and deliver the baskets, and arrive at a price that is appropriately outrageous. Sell them for Christmas, as gifts for employees, or whatever someone wants. Hang photos of someone receiving one where they can be easily seen, along with the newspaper article on the event, a handwritten note proclaiming them the best ever gift received, and/or a thank-you for producing it.

Your candle display should be a showstopper for your customers who don't do bees.

Pollen and Propolis

First, a disclaimer. Pollen and propolis have become problematic recently, and it is recommended you use care when collecting or selling either product. From some sources, these products have been found to contain minute amounts of antibiotics from beekeeper-applied medications, and pesticides from beekeeper and agricultural applications. If you intend to sell these products to retail outlets, or even to your customers, you may be asked to guarantee their purity. That can be done, certainly, but the cost can be an issue, and you may need to do it for every batch you collect. And changing agricultural practices will require continuous monitoring. You certainly have control over what you put in your hives, so you can be assured that your products won't have these compounds found in them. And if you exercise care in choosing beeyards, you can come close to eliminating any chance of agricultural pesticide residue being found in your products. But these chemicals are out there, and sometimes they will find their way into your world.

Of course, this raises a flag for what you feed back to your own bees from what you collect. If your bees are thriving this season and you are collecting what they have found, you can be reasonably certain that it is safe and clean. But season-long observations should be practiced, because problems don't always show up right away, and when you have one—queen loss, brood reduction, or any of the other issues contaminated pollen has caused—your collected pollen could be suspect. Collect and use with care.

POLLEN

You trap pollen, right? Probably not. It's amazing how few beekeepers trap pollen to sell, consume at home, or feed back to their own bees. What better food can you find to keep your bees as healthy as possible?

But now that you have to have your bees make every dollar possible, you can no longer ignore it. And yes, it can be time consuming and expensive, in terms of the labor it takes to trap, process, package, store, and sell it. But the profits are there, if you want to invest.

You trap pollen, right?

It starts with traps. Which one is best? Or, which one is cheapest? Or, which one is easiest? There are real differences in quality, ease of use, and especially cost. And it depends on what you are using it for. In a nutshell, here's what I've found: For pollen you are going to sell to the public, get the best traps you can find. They will keep collected pollen clean, dry, and safe. The best ones can be used on pallets with entrances on either side, and the drawers are easy to empty. This collected product may need very little, if any, cleaning, and the cleaning is usually pretty fast and easy. There's labor involved, but if you have enough bees to have pallets, this is the way to go. One stop, four traps. On for a couple days, off for twice as long or longer. Collected, home, cleaned, bottled, frozen. Done.

If you're collecting pollen to use for your bees, get the cheapest, fastest, easiest-to-use plastic models you can find. Your bees don't care if it gets wet as long as it's collected often enough not to spoil. A workable arrangement in a beeyard with six to eight pallets is to have only one, and at most, two of the good traps, maybe twice that with the cheap units. You'll find, after a bit of practice, that it takes you about three times as long to collect and dump the good stuff (which isn't all that long, really), but it's three times easier to work with than the cheap units that collect the working day stuff.

The cheap stuff goes into the freezer right away and stays until needed or until you want to work with it a bit for feeding later. You can go several ways with the good stuff. Clean it right away if it's dry enough, and if you collect it routinely it should be. Use a cleaner, or even just a small fan to blow away the lighter material. I've found using a small hand-held personal fan works; with the collected pollen spread out on a table on newspaper, move the fan at about a 45-degree angle starting on your side of the table and gradually moving away from you, blowing the debris onto the floor on the other side of the table. Practice will show you how close to keep the fan so it doesn't move too much pollen off the table along with the debris that was there. Don't spend a lot of time with this. Leaving some debris is okay, as long as any bee parts are moved. Losing a fraction of the pollen is both inevitable and acceptable. What's the value of a pollen pellet in time spent cleaning?

With good traps, cleaning may not be necessary, and you can collect and refrigerate or freeze immediately. Most use airtight sandwich bags right in the beeyard. Stick the bag in a cooler with some ice for transport home, and then keep it refrigerated until bagged or bottled for sale.

You can dry your collected pollen using any type of oven with pellets spread on a pan, but be extremely careful not to go overboard. Aim for "barely dry" as opposed to "very dry." You want them dry enough that you can put them in a jar or bag and they will pour out and not stick to each other. Of course, pollen dryers do a good job, and if you are collecting pollen by the bushel, they will be cost effective. Plus, some models can be used as a honey warming cabinet, too. Your label should be similar to your honey label but have a different color plan. On the first/top line, put "Bee-Collected" in medium-sized type. On the second, centered below the first line in much larger type, put "POLLEN." Include the weight. Use a jar unlike your honey jars so it gets noticed when on display. Know the weight of the pollen your jar holds and label it just a tad less because of the variability of pellet size and moisture

A gentle blower, a moving tray, and sharp eyes can handle lots of pollen cleaning.

content. That way, your product may be heavier than the label states, but it won't be lighter. State that it is honey bee–collected from thousands of local wildflowers, and that it may contain trace amounts of extraneous plant residue, because it might. Make no health claims, and be sure to say something about allergic reactions if ingesting large amounts. Suggest starting with no more than a ¼ teaspoon, or even less, on a bowl of cereal and testing slowly.

You can check your competition's pricing, but know the initial cost of your traps, the time involved in collection, cleaning, and preparing for sale, the cost of your labels and containers, and delivery time. A rule of thumb is twice the price of honey by weight, but don't assume that works for you.

PROPOLIS

To be honest, fussing with propolis any more than using a trap, freezing it, and selling the chunks you collect is more trouble than it's worth. But you may find there's a market for additional products made with propolis where you are, so don't ignore it. But do keep track of how much work it can be, and remember that your time is money.

Propolis is approximately 45–55% resins, 25–35% waxes and fatty acids, 10% essential oils and aromatic compounds (phenolics)—which includes vanillin, giving propolis its wonderful vanilla-like smell—and usually a tiny bit of pollen. There are also trace amounts of other organic compounds and minor components that researchers have yet to identify.

A propolis trap consists of a thin plastic sheet with narrow slits cut into it; it's the size of your inner cover. Set it on the top bars, lift the cover a tad to let in some light, and over a short time bees will fill the narrow slits in the plastic with propolis, sealing the hive from light, air flow, and unwanted visitors. Once the trap is plugged up with propolis, put it in a bag

and place it in a freezer. After a couple hours, remove it from the freezer and bang it against a hard surface such as a table. The brittle propolis will break away from the trap, and you can collect all the pieces in the bag. Now you have a collection of raw propolis.

You can simply sell these chunks by the ounce (or gram) in a bag or jar, pricing a given weight for however much you can get for it. You can also add some value to it by freezing it and grinding it into a powder for people who will use it for tinctures and the like. You can even make the tinctures yourself with alcohol, oil, or even water.

There are a lot of claims about how to use propolis medicinally; they range from holding a chunk in your mouth for a sore throat and gums to skin applications and more. And now you're getting into the gray area of bee products and medicine. Unless you have lots of experience with this, it's a good idea to simply supply those who do have more experience with the raw product and let them make the salves, tinctures, and such.

It all starts here, if you don't use a trap.

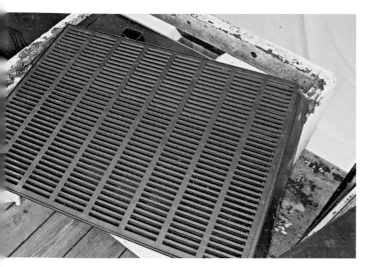

Propolis trap.

A propolis drum.

Propolis can contain beekeeper-applied pesticides, and even chemicals applied to the buds of the trees the original resins were collected from. You may run into issues with selling this product for human consumption. A cautious approach is to simply sell raw, honey bee–produced propolis, make no claims, give no advice, and leave it at that. Prices tend toward about eight times the current price of honey, but it's eight times harder to get it, so that balances out.

However, there is another side to this that is remarkably interesting and certainly worth investigating. In New Zealand, companies that deal with propolis—mostly the large honey packers—take a whole different approach on collection and processing. They supply beekeepers with a standard plastic propolis trap with a known weight. The beekeepers put them on their hives, prop the cover up to let in some light and let the bees do their thing and fill in all the openings to kill the light coming in from the top. When complete, or when the beekeeper decides it is done, the trap is removed and taken back to the company and the beekeeper collects a fee for the propolis on the trap.

The simplicity of this is incredible. The ease of collection is easy on the bees and the beekeeper and the beekeeper doesn't have to deal with any of the issues of processing the propolis or trying to find enough of it to make it worth the time to collect and

process. The company that collects the traps gets rid of the variability of the quality of propolis collected, with no worries about antibiotics (antibiotics are not used) or other contaminants in the finished product. Not that miticides aren't used, but like honey, they are not used when a propolis trap is on the colony.

Once collected, the traps can be frozen and the propolis banged out of the trap, or, the entire trap can be soaked in a solvent to remove the propolis and make it useable for the next step. Straining is needed but because there are no hive scrapings in the collection, no wooden splinters, seldom even dead bees, and usually very little wax, the resulting solvent solution is ready for the next step. That could be any of the many medicinal products propolis is supposed to be good for, and the quality of these finished products is excellent.

Would you consider this as a way to work in the world of propolis? Propolis traps are inexpensive and if acquired in bulk could probably be nearly free. To make this work you would need as many traps as you could find beekeepers to use. You could sell them to the beekeepers at cost or you could simply loan them to the beekeepers.

Like honey, they can only be used when there are no chemicals being used in the hive (other than carbohydrates), and the opportunity to have samples tested should be known. You can derive a price from local buyers, the internet, or the people who will be buying the harvested product. You can invest as little or as much into producing a finished product. Be aware that there will be some amount of wax in these, sometimes, so you can gauge the price paid knowing that there will probably be some weight loss in the end—estimate about 5 percent. But you can figure more exactly with your customers, factoring in their bees, the weather, and so forth.

By itself, propolis can be a headache, but dealing in bulk, with other beekeepers, and a buyer for all this bulk product can turn it into a profitable money maker. Don't overlook the possibilities.

"Bee Day"

If you make any retail sales at all, you absolutely need to organize a "Bee Day" at your place of business. Heavily promote it on social media and in local papers months in advance and get as many local, and even not-so-local, organizations and businesses to take part. It's your opportunity to help everybody find out more about honey bees, beekeepers, and beekeeping.

In the best of all scenarios, you will have all of what's outlined below. The space, your budget, and available personnel may limit you some, but your efforts will be rewarded with new customers, additional income down the road, and a grateful—and now more knowledgeable—community of both beekeepers and people who just want to know more. Some have tied an event like this to package day, which might work if you have good control over when packages arrive. If you don't—and spring weather is tricky—simply make it a "Honey Bee Appreciation Day" at Mike's Bee Supply Store. Bring the whole family!!

Set-up strategies—Start with tents outside your store. Big tents. Very big tents. Set up at least two and have them connect at a 90-degree angle. On one end, position the free food and drinks, and on the other side feature some displays of antique cars, beekeeping equipment, and so on. Fill the rest of the space (and outside the tent) with tables and chairs for people to sit and relax and nibble on their goodies from the food trailers on the outside and your free snacks and beverages on the inside.

Entertainment ideas—Somewhere in the middle you should have music on a small stage—anything from a small county music trio to a larger, louder group. If you have the room, make a u shape with three tents and use the space in the middle to host face painting, balloon hats for the kids, and extra tables and chairs for parents to sit and watch.

Vendor considerations—Set up tables for vendors along the perimeter of the tents, and, if you did it right and have room, you should have some vendor tables set up in the center so you have two isles in each tent. Position your largest vendors—probably you and maybe the other bee supply company you invited—at the end of the row. And note that you have to have at least one other bee supply company in attendance.

All your vendors, especially your largest ones, should be easily accessible for customers and protected from the elements.

What vendors should you include? Here are some thoughts; Reach out and touch as many of these folks as you can. You can charge a nominal fee to those who are selling things if you think it's necessary, but only cover the cost of their table and backdrop rental. (This reminds me, make sure you have something behind your vendor tables to protect from too much wind and light.) Start with community people. Local police and fire departments should be invited along with the local library showing their collection of beekeeping and gardening books to loan out. Have a nursery there selling native bee plants, and local crafters selling candles, lotions and potions, and knick-knacks for gifts. The local quilting society may be coaxed to raffle off one of their works with donations going to the fire department or library.

Be sure to have your local beekeeping association there. Perhaps they could be holding demonstrations way out back, showing off their four or five observation hives and talking bees with anybody who stops. Honey sellers must be there. And, yes, they are competitors, but you're not selling the exact same things, and therefore giving customers additional choices. Arrange for a cooking demonstration, but consult local regulations for such first. The demo can be ongoing, using honey and honey-bee pollinated crops. Also, vendors selling t-shirts and hats should be welcome, too.

Education and advocacy possibilities—Invite speakers for all the beekeepers that will be there because you've been selling this all winter as the place to be this spring. Off to the side, your beekeeping demos should be ongoing all day, but in at least two other places, set up a podium, computer and projector, and sound system and find people who can run them. Bring in a local big name, or a more distant big name, to attract the beekeeping crowd. Worked right, that same speaker could give a different talk the next night to your local beekeeping group and both of you might be able to swing a price break on the honorarium. Topics should include very basic beekeeping, a bit more advanced beekeeping, and something special, such as a seasonal talk on summer management, or installing packages. If you have the room, convince the local master gardeners to speak about producing pollinator gardens and other honey plants. But don't forget to treat your speakers right: don't overschedule them, make sure they have food, water, shade, and maybe even a fan if it's hot.

Devilish details—If you sell more than bee supplies at your store you should have a separate table (or more) for your "gifty" stuff, plus your honey, lotions, and potions. Chances are, if you group it all together in your bee booth they will get lost and nonbee people won't get to know what else you do for a living. Show off a little.

Have easels all over the place featuring the day's schedule, what speakers are speaking when and where, especially, and all the rest of what's going on.

Also, print a program for the day (with a coupon for a free honey stick at your table) to give to everyone who comes. Appoint someone to oversee parking all day long. You don't want accidents, and you do want order. Clearly designate a main entrance where people are collecting entrance fees, (Yes, charge at least $5 for an adult, but allow kids for free.) giving directions, and keeping a headcount. Also, have volunteers dressed in attractive, identical shirts with the name of your business on them, strolling around the area with water for the speakers and anybody working there and stopping to ask customers if they are finding what they need. Finally, you need restrooms. Rent a few, set them where you can just see them, and be sure to have signs everywhere pointing to your restrooms so nobody needs to ask.

Every one of your employees is working double duty that day, but you are not. Your job responsibilities should be checking on the speakers, talking to vendors, helping the chef, moving chairs to accommodate a bigger crowd, answering questions about your business, and actually having a grand time with all this. "Bee Day" only comes once a year. Enjoy the day, and reward your employees for all their hard work. It will be worth the cost, labor, and aggravation. Your community will let you know that in spades.

Invite local charity organizations as well as your local police and fire departments.

Sell your honey separate from your supplies, and make sure other honey sellers are included.

If you have entertainment make sure there are places for people to stay and enjoy it.

Having the appropriate sound system and seating on hand

. . .is as important as your roster of speakers.

3 What Else to Sell— Services by Bees and Beekeepers

If you haven't carefully considered what your business will be doing, start now, before you make any final decisions on your facilities, storage space, and the like. It's time to review what you are actually going to be doing that you haven't been doing already.

Right off, you've probably already figured out how much time you will spend taking care of your bees. If honey is the big player for you, you may prioritize time spent working in the store differently than if honey is only a minor player or you have someone else running the store full or part time. So what will you sell besides honey?

Because you've already studied your potential customer base, their buying habits, and your competition when completing your business plan, you have a fair idea of what equipment and materials customers will need and want. You'll know if you have to sell pallets, boxes, or single items, which tells you how much room you'll need in the store and the warehouse.

The bigger question is not what you will sell, but where you will get it. Becoming a dealer for one of the large manufacturers is an obvious option. The qualifications you need for most of them are similar, so it comes down to the choices they offer, the quality of their products, and the interchangeability of the equipment they sell. Much of what everybody sells is similar, so price, minimum purchases, and shipping costs may be the deciding factors. Get all the details from each manufacturer you want to deal with:

- What are the discounts?

- Do shipping costs vary by amount ordered?

- Is credit a possibility?

- Is becoming an exclusive dealer a possibility?

- Is there a minimum purchase per order and per year required to be able to say you are a dealer?

- Is there another dealer nearby and within however many miles?

- Can you be a dealer for more than one manufacturer?

- What sort of facilities do they require?

- Do you have to sign a contract?

- Will they help with advertising costs and customer support?

- Will you be listed in their catalogue?

- What will they do about returns, damaged items, and billing errors?

- Some businesses like yours get everything but, say, wood from one supplier, but they get all their wood from another supplier because of quality. How does that work? Does it work?

What else? Being a dealer has lots of advantages for your business. Choosing the right manufacturer or two won't make or break your operation, but the best one will help more than the others for reasons that may be as simple as freight costs or how well they support their dealers.

A significant consideration when working with more than one supplier is if the equipment is interchangeable. And, is it consistent over time? Sometimes, your supplier may have a quality problem in which their equipment this time doesn't fit their equipment from last time. Warp happens, wrong measurements happen, and suddenly you have a pallet of stuff that's only good for firewood. Don't for a minute make the assumption that it's all the same every time. It won't be, and if you do, your customers will not be happy with what you are selling. There's an old saying about measuring twice and cutting once. When buying already cut wood, measure twice and never have your customers return rotten wood. That's a good thing.

If using more than one supplier, make sure all the parts and pieces from each fit together—frame spacing, boxes, and the rest.

You can go independent on some items, such as Chinese smokers, which are going to be cheaper than anything you can find locally made. (Check, because that may have changed by the time you read this. Who knows where anybody is sourcing anything anymore!) Plus, you may choose to carry two levels of quality, the good, expensive stuff and the cheap, inexpensive stuff, so you can cater to both levels of customers. Your normal supplier is doing some of this, too, to keep costs down and remain competitive with Amazon. In fact, Amazon is one of your most serious competitors. If you haven't yet, check them out and see what you, and your suppliers, are up against. The products from China—and to be fair other countries, too—look identical to lots of products made here for two or three times the price.

Unfortunately, the same can be said for honey, pollen, and propolis. Your advantage, in every situation, is that you can answer customer questions.

By now, you know everything you need. And if you aren't sure, look at every bee supply catalog and website you can find and start deciding. The biggest factor is inventory. How many of every kind of bee suit can you afford to have on hand all of the time? And how many assembled, painted deep supers can you afford to have, even if you did the assembly and painting? Your business plan will go a long way in helping you decide because it pretty much sets the limit on how much you will be spending on this, at least initially. The old saying about avoiding debt is worth its weight in gold here, without doubt.

Liquid Feed and Protein

You'll need to decide what bulk liquid feed you'll carry. Will you have high-fructose corn syrup (HFCS) or sugar syrup for those customers who use enough to buy rather than mix their own? Will you sell it in their containers or already pailed? Finding a supplier is a matter of simple research, and the tanks and hoses and valves are usually available locally. Consider investing in security with some type of lock or other device to make sure it all stays where it is supposed to stay.

But this brings up a question that you will revisit again and again: Will you sell something that some beekeepers or others consider bad for bees? HFCS has been accused of being unhealthy for bees. Some feel that many miticides are unhealthy for bees, wax, and honey. Do you choose to sell them anyway? You probably won't lose beekeeping customers, but you probably will lose some sales to beekeepers. From where I sit, as Mark Winston would say, it's an easy choice. You have to sleep at night; you have to have a clear conscience about what you are recommending to your customers. If you are selling it, you are recommending it.

The many forms of protein available defy categorization. Which is best? Your customers will tell you eventually. Have two on hand if you can, preferably three. If you can manage it, sell each at a premium

You can sell protein as powder, as bulk patties such as these, or prepackaged in boxes.

price. Most hobby folks won't need a box of twenty units or more; maybe they'll just need half a box. Make it easy for them. At the same time, if someone you know will use a pallet, can you get it and can you load it? And keep up with the new stuff coming on line. Purina Bee Chow is real, and you have to know about it.

If you are an exclusive outlet for your dealer, stocking more than one protein may be an issue. But here's a trick question: what do you use? If not your supplier's, why not? Be ready to answer that. Maybe it's not the best, or it's the most expensive, or . . . well, someone is going to ask.

Sugar syrup truck for delivery.

HFCS storage tanks.

Indoor tank for small quantities.

BULK DISCOUNTS

Will you provide bulk sale discounts routinely, or only when it works to your advantage? Let's say a customer, probably a nearly large sideline or commercial customer, wants five or six pallets of knocked-down deeps from not-top-quality wood for his help or his family to assemble over the winter for next spring's splits. With your connections, you can save him quite a bit because there's a truck coming from your wood people anyway, so it's essentially no additional freight cost for you. That saves a healthy chunk of change for that customer. Remember, it's a small world you are a part of. With what seems like no additional cost to you, you can do a major favor for a customer—or someone who may become a good customer. Easy choice, right?

Maybe. Look carefully.

Pallet loader issues could arise. You have a jack for the warehouse, but say his pickup doesn't match your loading dock, and it all has to be hand loaded. By his crew, your crew, a mixed crew? Time is money, and you didn't ask beforehand. Then he starts looking and rejecting this super for having too many knot holes, then another because it's split. Then, that pallet looks smaller than the rest; is the count right? The deal was for a shrink-wrapped pallet, loaded and gone, not hand unloaded and picky. Be careful, you could end up losing money in labor and time on this favor.

Woodenware Suppliers

Can you have some of your woodenware made locally? You may be in a part of the country that supports a lot of woodworking businesses. If so, contracting out supers, covers, frames, and the rest may be a no-brainer because of cost, freight, type of wood, and especially quality of finished product.

MAKING SOME OF WHAT YOU SELL

Some people start keeping bees in the first place because they get to use their woodworking skills to make their equipment. Maybe you're one of them. Maybe until now, you've been able to keep up with the demands of your beekeeping operation. But will you have the time to produce twice or ten times the amount of woodenware you were making? And did you calculate what the actual cost of one of those supers is? When you are buying lumber for, say, twenty supers, you are paying more per unit than when you buy enough for a thousand. Can you purchase lumber competitively, add in your time, sit on inventory, and make the profit you need? Maybe, maybe not, but you won't know until you *do the math*.

The other issue with homemade wood is that it has to conform to other equipment. Do your supers and the frames from your supplier play well together? If not, do you find another supplier? These are simple questions, but you need to ask them before you invest in more equipment and woodshop space.

Is your wood supplier's quality and fit consistent all of the time?

WOOD FROM YOUR SUPPLIER

So if you're not going with homemade woodenware and there isn't anybody local to make it, what then? If you're a dealer, how good is the stuff your supplier makes? Good enough? Almost good enough? Too expensive? Mostly budget quality? Putting all your eggs in one basket is one way to go, but it has its dangers. Then again, it's also dangerous to get frames from here, boxes from there, and inner covers from somewhere else. It all has to fit, and some manufacturers don't have consistent quality due to labor issues, lumber suppliers, and the like, no matter how hard they try. So you have to have a plan B for all this. You have to find out whose equipment fits whose perfectly so you don't wind up not having what your customers want or you want to sell. But really, is there that much difference? You don't know for some time usually, and then your customer comes back with a complaint, maybe even brings in the burr comb-covered inner cover and asks what's going on. And then you'll have to ask some questions: Did you buy it here? How long have you had it? How long did it sit unused? Where did you get the frames and the box below that inner cover? You are in the business of selling beekeeping equipment that works, and suddenly, through no fault of your own, it doesn't. What then? Have a policy on hand because it will happen—not often, but it will.

One way to avoid any fitting issues is to make it yourself. Does that work for you?

And then there's budget wood . . . which comes with no guarantees, ever. But it's cheaper and useable and good enough for some operations. Definitely have a line of budget wood from somebody. You have customers that want only that stuff, so stock it. But know the issues. It's budget for a reason.

There aren't many other beekeeping items that can be made in the backroom. Bee suits, smokers, harvesting and processing equipment, anything else metal, chemicals, and the like all pretty much need a formal manufacturing facility. You could bundle smoker fuel or make specialized feeders from jars or cans, but demand for that is low. Mostly, it's going to be wooden boxes, frames, covers, top feeders, pollen traps, inner covers, hive stands, and the like. No matter what you make, make sure it's a good fit for all the other pieces of equipment your customers will buy from you or from somebody.

If you hate plastic equipment, will you sell it to your customers who love it

Business of Business

One of the initial costs we discussed was internet access in your store and cyber insurance. You have to take credit cards and debit cards, and you have to be able to get on the internet at your checkout. These aren't choices; they are givens. Your bank and your accountant will make this happen, there will be a cost, and you can't ignore it. Yes, some small businesses do not, will not, deal with this. But who are your customers, and who do you want your customers to be? If you're shooting for the geriatric set, it might work, but most folks don't shy away from technology, and they will expect you to take their debit card, credit card and maybe PayPal, too. And your website should handle all of these.

Your accountant, bank, and attorney will have set you up to collect taxes, to pay them, and to give tax breaks to some customers who qualify. That will all happen during the initial financial work, so it should be in place for the very first customer.

 Technology at the checkout counter. Will you accommodate Apple Pay and all the rest of the cell phone tap pay techniques that are available today (and even more tomorrow)? What's out there? Go to a local coffee chain and see what's going on. You have to be able to do the same thing.

Education for a Price

We started down this road when we discussed teaching classes on making soap, candles, and the like with beeswax. Holding classes at your store for non-beekeeping customers will build your base, increase demand for the raw materials and instruction books, probably increase honey sales, and help spread the word that your business not only exists, but is the only store like it anywhere close.

The same applies to beekeeping classes. Yes, the local association has classes every spring. But what else? A monthly meeting with a speaker maybe, a summer meeting in a beeyard, a Christmas supper, and that's usually it. You can offer anything not offered by the local group. And, run classes for people with different skill levels—beginner, intermediate, and advanced—plus specialized skills. Face it, you're a good enough beekeeper to teach almost anything a beekeeper needs to know, plus you know a host of other, non-local beekeepers who are experts in specialty subjects.

You have to have a place to do this. If your warehouse or store won't work for lectures, find a library, church basement, or even wherever the local group meets. You aren't taking business from them, and you need to stress that. The other classes can most likely be at your place; you have the bees and all the equipment you'll need. Limit the size of the class to fit the space you have.

After the beginner's class in the spring (and depending on the local group, you might even offer a follow-up beginner's class stressing what they don't), you can run classes such as Care for Your Package, Ready for the Honey Flow, Producing Varietal Honey, Harvesting Your Honey, Setting Up Your Extraction Room, Bottling Equipment, Label Design, Getting Ready for Winter, and Second Spring, all designed for that first-year beekeeper. Yes, it will require investment in prep time and class time, but the class times, at least, will be at night when students can attend and you aren't taking care of customers. You may even have to run two or three of each of class if there's enough demand. Charge a modest fee

Give outside hive inspection demos as often as you can.

for these; the local group charges for their beginner's class, and you should, too. And of course, offer the supplies you are demonstrating to your students at a discount, making their life easier and them better beekeepers.

You will hear from some of the old-timers in the group that the only reason you are doing this is to sell equipment. It's an ageless curse in the beekeeping industry that you've probably already heard. "Dealers are the worst scum in the universe," goes the chant, "greedy as all get-out. They are only after a poor beekeeper's money." It's older than A. I. Root, and it will probably never go away. When you hear it, and you will, you can always offer that old-timer the opportunity to help teach the class. You'll pay him to do it and offer him the same (or maybe

even a bigger) discount on the equipment he'll be using to teach. And make sure you have some of those other old-timers—and some of your potential students—within earshot. If he refuses, you've at least made an honest attempt to show that what you care about is helping beekeepers succeed. If he takes you up on it, you won't hear about what a scam it is anymore. It works.

Consider also offering advanced classes, for that old-timer and anybody else who wants to learn. You could run classes on making splits, rearing queens, making your own woodenware, pest and disease monitoring and control, what's new this year, winter wrapping, different feeding techniques, and using remote sensing equipment and analyzing the data collected. And, of course, you could run refresher courses on any of the beginner classes you've held. You may need to bring in an expert or two for these, and you'll have to pay them, sometimes dearly, for their skills. The tuition from the class should cover most of this, maybe all of it. Note that guessing the number of attendees when you double your normal price may not work. You won't know until you try.

If you work it right, you could offer some of this educational programming, jointly, with your local beekeeper's group. For example, let's say they bring in a queen producer for their monthly meeting. The next day, that person could be at your place—with queens, drones, cells, and all the rest—for a hands-on demonstration. The additional cost can be split between you and the group, or maybe they pay it all and get to use your place for the demo. Either way, you have folks at your beeyard you didn't have before. Suddenly you're the go-to guy for that skill, and you have all the equipment they need and the skills to show them how it works again.

Demonstrations can be tricky, but they can work. And if you're looking for something more basic, extracting honey is the typical show-off tour you can offer to groups, schoolkids, and the like.

LECTURES

If getting up in front of a crowd is something you kind of enjoy, you can use that skill to your advantage as a promotional activity. Starting with your local association, you can develop a reputation as a teacher on several subjects. Then you can reach out to non beekeeping groups to help them better understand bees and beekeeping. Local social and civic groups need speakers, and they might have some topics of interest in common with beekeepers: preserving pollinator habitat, using pesticides on lawns and gardens, what to grow to feed the bees, responsible mosquito control, commonsense good-neighbor beekeeping, improving regulations and guidelines for urban beekeepers, and the like. Some of these are universal and can be shared nearly anywhere.

You'll likely be doing this for free. It can be your way of giving back to the community, and it's an opportunity to let them get to know you and your interests. Being a member of any and all civic groups is also part of this. Yes, time is money, but sharing your skills is important, and perhaps you can use what you learn at these gatherings to better your business or make contacts—with someone who can help you with your problems in banking, legal, insurance, IT, or HR, for example.

TOURS

With bees and beehives, tours might be tricky, but they aren't impossible. With honey processing, they're definitely possible. If your facility lends itself to this sort of activity, don't avoid it. You'll need an extractor room with a big window into the rest of the honey house, and probably it can be temporary by blocking off some part of the room with windows or such. School classes can visit and watch the whole process of unloading, uncapping, extracting, and even bottling. You can charge a small fee for the class to come, get a short lecture on what's going on, and go home with a couple small souvenirs—a honey stick or two and a beeswax birthday candle or orna-

Show off your honey house, even giving demos, if possible.

ment—each. No, you won't get them as customers, at least not for a while, but they'll have a good memory of you and the bees to tell their parents about it, and their parents might bring them back. And, in the long run, you'll have a generation of people who know more about bees and beekeeping than their parents did. And that's a good thing.

LOCAL LAW ENFORCEMENT RESOURCES

If you have the equipment and skills, bee removal can fill odd hours, serve as a gift to your community, and get you free bees. Look into it, but if it would be a time drain for you, consider training a few local beekeepers on the basics needed to remove bees from troublesome places. As a local beekeeping resource, you will become the go-to person that anyone, including the police and fire departments, will call when bee removal is needed. Even if it isn't you pulling off the siding, somebody will be happy you knew who could do it. Swarm calls go somewhere, and they'll call you because you always seem to know who can get that swarm off the storefront on Main Street.

TIP: Develop an emergency plan. Working with local police and fire departments, develop an emergency plan for that worst-of-all accidents that could happen. Figure an overturned semi in the middle of the night that is blocking a freeway entrance. Do your local people have protective gear available, do they have veils, suits, and the like so when they are standing out by the truck directing traffic they are protected? These should be available in every fire truck and fire rescue vehicle. Do police have the names and numbers of local beekeepers available to dispatchers to contact immediately, and do those contacts know what to do when they get the call? If you can, actually stage an accident. Show how forklifts can help, what fire departments can do with simply water, and what water with wetting agents can do to control a life-threatening situation. Who is in charge? A few classes and a couple of field demonstrations will put this in place. And then a refresher held every couple of years, as technology and new people change the landscape, will go a long way in making you the go-to person for emergencies and the person responsible for saving lives.

Where can you get a forklift in the middle of the night, let alone someone who will use it to move a tangled mess of bees after an accident?

That police and fire department connection can manifest in other ways. If your operation is adjacent to major bee transport roads, an exit ramp roll-over is inevitable. State, county, and local authorities need to be able to react quickly to minimize road closure and maximize public safety. Coordinating a rescue like this takes advance planning and communication. Where would you go at midnight to get a forklift to help reload a truck? Who knows ten beekeepers who can react in minutes rather than hours, or even days? Helping local law enforcement develop a plan to confront problems goes a long way toward establishing good relations with the community, the local beekeepers, and the companies that own the bees and the truck.

A demonstration on preparing to move bees is popular because everybody has to move bees sometime.

Take this to another level. Do a practice run. Know that the forklift is available and that it works. Find the beekeepers who will come at midnight, and be certain the police will keep you safe when you are reloading that truck. Do they have protective gear? A good bet is that they don't, but they should. You can help with that. Know that the fire department will arrive, in a hurry, with a truck with surfactant if you can't get the rolled truck unloaded, and that they have safety gear. No, you probably won't make any money on this one, but you may save lives because everybody knew what to do.

FIELD DAYS

Having at least one free field day every spring or summer is a given. You simply can't skip this. Bring in both local and distant speakers. If you don't have room inside for the crowd, you'll have get a tent; figure out the audio ahead of time. Parking can be an issue, so figure it out far in advance, too. Have something for kids so their folks feel good about bringing them, and so their folks will come in the first place. Have open hive demonstrations going on all day in several locations, showing very basic to very

Everybody has to have protective gear. Your insurance company will appreciate that.

Put up a tent in case it's hot.

advanced skills folks can watch. Vary the demonstrations a bit every year so there's always something new to see. Insist on protective gear when people are watching a demo, but have some on hand for those who forgot theirs; it can be simply inexpensive mosquito veils. Use all the talent from the local group you can muster so they get to display their skills. In addition, invite people from other groups so everyone can meet and greet and perhaps work together, later. You're not on display today. It's not you folks are going to come to watch and not you as the center of attention. You simply organize it all, keep it running, put out the fires that will arise, and make sure everybody has a good time.

Prepare free food or pay some local church or other group to come and prepare it for you. Perhaps a local food truck could do the preparation part and you cover the cost. Drinks on hand all day in tubs of ice should be part of the deal.

And yes, your store should be open. And you should have more than enough help inside and at the special equipment displays you put outside to answer questions and demonstrate. You can have specials, equipment you don't normally have, and maybe even a representative from one of the specialty equipment manufacturers or from the company you are a dealer for. Maybe a queen and package rep can be on hand to show-and-tell, too.

A day like this is an investment in your future. It can be an expensive day, and you need to make sure it will pay for itself in the long run. And you probably can't do all of this right off, but you can do some of it. If you start small the first few years and gradually add in something new every year, you'll build an event-like expectation for those who attend every year, and you'll have a reputation for all those new folks.

If you have room, build a fence like this: bees and beekeeping on one side, viewers up close and personal but protected on the other. It's 8 feet (2.4 m) tall, and the viewing area is covered in window screen. You can see, hear, and smell the bees.

Do a bee cross. Fasten the queen cage to the center of the cross with her bees in a nuc early in morning. The bees will move up to cover her. For the demo, move the cage back to the nuc, or to another cross, explaining the reasons the bees are moving, talking about pheromones, and explaining as you go why you don't need a veil.

Provide activities for kids, such as trying on bee suits.

Don't forget "Johnny Bee Good" in his bee suit, handing out honey sticks.

Where Else to Sell

You can sell the things you buy and make in places other than your store. Beekeepers are outside people, so why aren't outlets that cater to outside people dealing with beekeeping supplies? Take a quick look at any seed store, hardware store, or even some nurseries today, and in the back there's a shelf with beekeeping equipment on it. It's a lot more retail than what you are used to seeing—fancy boxes, labels, and such, which are what normal people expect to see. (Okay, beekeepers are mostly normal, but they aren't really used to this kind of fancy.) Can you compete with this? Yes. Here's how.

Stress local. Really stress local. Your label should scream "Made in [YOUR County]." And what are you selling? Knocked down boxes in a fancy carton? Hardly! You've got assembled, painted supers. You've got hive stands, covers, inner covers, smokers with a box of fuel, feeders, and all the other equipment a new beekeeper will need. It's not for every store, but try as many as you can get into. Make sure the pricing is such that you at least break even and that the prices are competitive with yours, the internet, and the catalogue of your supplier. Your hook is, when the customer buys that super and comes back with a question, where is he sent? To you. They start at the feed store, but they end up in your store. The feed store knows it, but because beekeeping isn't their specialty, they don't care. They make their buck and move on, and they don't have to figure out the answers to all those questions. For you, it's a win-win. In fact, even if that customer buys the competition's super or bee suit in the fancy box, the store will still send them to you when they have questions. But if they don't know you exist, they won't send them to you. Your job is to make sure they know you exist.

Find out what they don't have and supply those items. Especially important advice: Put your name and number on every box on the shelf.

TIP: Take working with local farm stores another step. Offer classes and field days to people from the store. Actually teach them how to teach. They may be a competitor when it comes to selling packages, but you can make them an ally when it comes to selling the right medications, the right kinds of feeding supplements, and so on. When push comes to shove, they will send their customers to you for additional advice, additional equipment, and better equipment. They will get people into beekeeping, but you will keep them there. Work with these folks, not against them. It will pay off in the long run.

Think Local

We talk about the advantages of flaunting local, but it works both ways. Everything you sell should advertise where it comes from. "Locally made" is your mantra for everything you can produce yourself. Use it. But how do you get the word out? You work within your community by raising your hand to serve, whether it's through church, sports, volunteer social groups, or civic groups. Anything that serves the rest of the community is recommended, because every time you attend a meeting or event you have the opportunity to promote your products. Not, of course, with gusto; be subtle with a monogrammed shirt, a pen you leave behind, a sample basket as a door prize, or any of the other little ways you can keep your name and business in the good light of giving. It truly does work both ways. Everybody gives a helping hand, and everybody gets the nod of recognition.

This works especially well with honey. If you can convince your local school, athletic club, university, restaurant, food manufacturer, or church group to use your locally produced honey, they have something to brag about—that is, they buy local and support their community. And you have a name-brand local customer; "Medina State College uses our honey because they recognize quality!"

Working with your local association, sponsor an award in the local honey show.

ASSOCIATIONS

You should hold a worker bee position in as many beekeeping and gardening associations as you can find and have time for. Yes, there aren't many, I know. But there are some. Tip: Don't be the president; be the set-up person or the coffee maker. Be valuable, but don't be too visible. Don't get caught making decisions or, sure enough, that old-timer will light up again about you only looking out for yourself. You can work with the program person on finding speakers because you know every speaker in the five-state area. They'll look good and your group will do better because of your connections. And you don't have to take the credit. This goes for field days, too. If your club has its own beeyard, you can take a step back, but if they don't, you have both bees and liability insurance (something most members probably don't have), and you can offer to help. Let someone else do the demonstration or show off what they know. You just offer parking and maybe a few extra veils for those beginners who don't bring their own. You're not the president, not the decision maker. But be there when they need you.

WORD OF MOUTH

All of this—civic membership, law enforcement work, association assistance—of course generates word-of-mouth advertising, the most positive and cheapest form of advertising you can get. This starts with your in-store promotions, farmers' market retail sales, and the promotions you do with civic groups. Anytime you and your honey get to be in the same place at the same time, you can make note of people's comments, encourage them to tell their friends about the best honey they ever had, and leave some behind. After all, when people see that it's your honey being used here, they'll know it must be good. Sample tastes really go a long way here (see page 63).

LOYALTY REWARDS

Many retail outlets, no matter what they sell, don't know their customers at all. But how many of the places you shop have a loyalty program? You know, that little card you carry in your wallet that gets scanned at the cash register? Does your store have a program like that? And what do your customers get if they are loyal? They want money, prizes, attention, free stuff—things that everybody likes. Your loyalty card can be as simple as a punch card, which puts the onus on the customer and not you. When it's been punched enough times, they get something for free. A barcode scanner, if you have one, is easier than punching. If your IT people are good, loyalty rewards can be hinged on a phone number. Enter the phone number, and the software keeps track of the number of visits, the total purchases, or whatever you are tracking.

Once you have a loyalty-rewards program set up, advertise it. Have a sign by the cash register. Put it in your ad in the local association newsletter every time; it will remind the regulars and catch the new members.

You can create loyalty by offering a new customer a free double dip the first time, giving them a leg up on that freebie you are offering, just because you're nice. Don't miss the opportunity to let this work for you, especially with new customers who will keep coming back.

GIFT BASKETS

One way to share some promotion with another local business? Pair with them for a gift basket campaign. Fill a beautiful basket with flower seeds, maple syrup, candles, lotions and potions, local cheese or beer, several kinds of honey, teas and coffees, soaps, honey sticks, cheeses, a mug, and/or a jug of something good. Of course, you can fill a basket yourself, but pairing with another local supplier gives a gift giver twice the reason to use your baskets rather than something not local or not as good. Seasonal gifts are a great option here: honey and seeds in the spring, honey and tea in the winter, or honey and candles at the holidays. You can always have some of these in inventory, as well as a lot of photos—in fact, a whole book of photos to show customers if they want to special order one. This keeps your inventory costs near zero, yet your product selection in full view.

Taking this a step further, you could sell these to other businesses to resell in their stores or catalogues. For someone local, you supply different sizes, different objects in the basket, and different baskets. For nonlocal businesses, you can work together. Find an attractive basket or box that fits three or four items snuggly, with a bit of a tissue paper cushion for starters. Then make up local baskets; for example, Ohio Wildflower Honey and Cheddar Cheese Baskets, Maine Wildflower Honey and Wild Blueberry Jam, or California Sage Honey and Smoked Almonds. All of these are local to those states, and the components don't have to be that close to home to count as local, for these purposes. Best of all, you don't have a lot of inventory to deal with because you just buy it in, put it together, and send it out.

"Gifty" items for both beekeepers and other shoppers are always appreciated . . .

. . . and they can either be related to bees or not.

SWAP MEETS

Something we don't see nearly often enough is a good swap meet. Try working with two or three local groups to hold one. Register attendees ahead of time so you know how much space will be needed. Each attendee brings a table and whatever equipment they want to sell, swap, or give away. Then let the games begin! The tables will be filled with extra smokers, slightly used bee suits that don't fit their owners, ten-frame boxes after someone switched to eight-framers, queen rearing or feeding equipment that never gets used, swarm catchers—anything and everything anybody has that they want to get rid of or trade for. Your store will also be open, and you'll have some specials and outside displays, but you probably won't do a lot of selling that day. However, you will get people to your place, and if you promote it right, a lot of them will be brand new to your store. They'll drive further to try somewhere new in order to get equipment for cheaper than anywhere else.

The only caveat is that used wooden equipment can be contaminated with disease. You could allow new, unused wood, or wood with a current clean bill of health from a local inspector. Or, maybe just don't allow wood at all.

Your Basic Catalogue

Let's take a quick look at the basics you will probably be carrying. We've looked at woodenware and bees pretty closely, and we've talked about propolis and pollen. But does your supplier make or carry all of the products, plus more, on the list below? Are they up to your standards? If not, you will have to find them someplace else. Of course, you don't have to have all of these on day one, but a beekeeper who walks in on day one is going to have some expectations.

- Protective gear
- Several hive tool choices
- Different styles and sizes of smokers
- Live traps
- Harvesting and processing equipment
- Uncapping tanks in plastic and metal
- Extractors in different capacities
- Strainers in different mesh
- Tanks in plastic and metal, heated and unheated
- Refractometers
- Totes, barrels, and pails
- Glass and plastic jars
- Labels
- Observation hives
- Hive stands, covers, boxes, feeders, inner covers, frames, foundation, and everything else it takes to make up basic five-, eight- and ten-frame equipment
- Styrofoam equipment
- Books
- Honey plant seeds
- Decorative skeps
- Gifts and gift baskets (see page 93)
- Beginner's kits. Your supplier will probably have several kinds of beginner's kits available already prepackaged and ready to go, and seasonally you'll need several in stock. However, during the off season, and even in the spring, you can make your own kits from existing stock. Include hive bodies, supers, smoker and fuel, bottom, top, feeders, frames, and anything else you would recommend.
- Craft supplies, including candle molds, lotion and potion additives, containers, and labels.
- Wax melters
- Wax and plastic foundation supplies
- Educational supplies: books, posters, teaching hives for classes, CDs, and the like for individual beekeepers and association meetings
- And what about those other kinds of hives? There hasn't been a standard top bar or Warre hive put on the market to speak of, so should you bother to have these? There are a couple of suppliers of the actual hives you could buy and resell. There are some pieces of management equipment to look at, including frame tools and feeders. See what your customers want and respond, rather guessing what they might want.

All kinds of protective gear: jackets, full suits, ventilated and regular, simple veils, helmets, and all kinds of gloves.

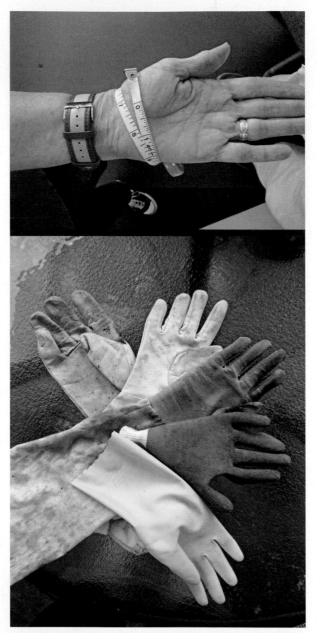

Various hive tools, frame grips, and frame rests. Include any other tools you use when working a hive.

Be sure to measure the beekeeper's hands so they get the right size gloves. The tape measure shown here, in inches, is the size of the glove that will fit well.

Seasonally, make sure you have what every level customer will need to harvest and process all kinds of honey, from fume boards to extractors to all sizes of storage tanks.

Make sure you have several styles, sizes, and kinds of smokers, and some packaged fuel to go with them.

TIP: Catalogues come in a couple of styles, basically. The first resembles a department store flyer in a newspaper. How many items can you get on a page is the goal—small photos (or no photos), limited copy for descriptions, and as few pages as possible but showing everything you have available and the price. The other style is just the opposite. You'll find very few items on a page, limited but stylistic copy, and maybe not even a price, which instead you will find on the order form. One thing to keep in mind is the cost of the real estate each page represents. An inexpensive item taking the same space as an expensive item doesn't make sense, so give each its due, but keep the cost/item-to-page-space ratio in mind.

If you can sell a tank, you can probably sell a better knife to go with it, along with better filters.

Filters for every need.

Extractors and other large pieces of processing equipment take a bite on inventory costs, and you won't sell lots of them. The smaller models will move faster, cost less, take up less space, and can be shipped easier. Order the larger pieces individually, but keep the smaller units on hand for impulse purchases and so your customers can take that next step.

Eventually, every beekeeper needs a heated bottling tank.

Keep in mind the storage space needed for cases of all kinds and various sizes of jars, bottles, and containers.

Containers for your largest customers, such as these barrels and totes, take up space, but there's no other way to deal with lots of honey.

It goes without saying, but you have to have the boxes and tops and bottoms of beehives.

Rigid, reinforced Styrofoam equipment is becoming popular and in demand. The new styles, with plastic ridges on the tops of the boxes to keep them together, work well, and mating nucs made of this material have proven themselves useful and durable.

If you can arrange it, set up several styles of observation hives, from the small single-frame educational size, up to as large as a three- or four-frame permanent type for schools or homes. You should always have one going in your store, and maybe a loaner for your local association.

Don't forget to have books on display and easy to find. Beginner and intermediate subjects will be the most popular, but specialty books on queen rearing, and making increase will do well, especially if customers can look through them.

Have appropriate honey plant seeds available, in season, especially in backyard-sized packages and larger. If you work with a seed company, packages can be designed especially for your area, with your business name on the package.

Think about having displays outside your store, maybe in the beeyard, by the front door, or near the entrance to your driveway. Skeps attract a lot of attention from both beekeepers and people interested in bees and bee products.

Gift baskets don't have to be static, and you can make them to order so you don't really have special inventory. They can feature honey, honey-containing foods, decorative objects, books, candles, lotions and potions, honey sticks, honey candy, and more. Advertise these for local businesses to give to employees or for associations to auction off as a donation. Collaborate with another business so you have an even wider selection.

Certainly you need solar wax melters, or at least the kits to make them, but have higher-powered models available that separate wax and honey, and others that simply melt wax.

You will have every kind of beekeeper as a customer: those who won't use any foundation, or even frames, those who are traditional and want pure beeswax foundation in all the right sizes, those who will use plastic snap-in foundation with wood frames, and those who want it easy and use one-piece plastic frames and foundation. If your supplier doesn't carry all of these, find sources for them, because you will sell all of them.

You will have top bar hive customers who still need food, medications, tools and harvest equipment, queens, and bees. Position one outside, easily visible.

Pollination

If you haven't explored this before now, it's time to look at the possibilities. Local contracts are often available for only a few colonies for season-long vegetable crops, spring fruit crops, and the like. These take little time and usually aren't very dangerous for your bees. And you probably already know the grower; if you haven't worked with him before when his usual beekeeper didn't have enough colonies for whatever reason, he made the first move when he found out about you.

Let's review the basics, because the world of commercial pollination is changing, like the world of commercial honey production, and it can only help to be aware of these changes.

First, and foremost, you need to know the crops your bees are going to pollinate better than the grower. No excuses. You need to know spray schedules and the chemicals and application techniques they will use. Is the number of colonies you will use the right number for the crop? You need to know about the people who will be in the field, the people who live next door, and where your bees will be in the fields. You need to know about a second crop they may want pollinated with your bees, what that entails, and how much you will get paid for it. You need to know approximate in and out dates and how much time you will have between that phone call and your night moves. You need to know irrigation schedules and where water pipes are so you don't run over them at night. And can you get in to inspect anytime you want? You want to make sure the bees don't get moved by anybody but you. Are there going to be any pipe ditches dug, buildings built, fences erected, lights installed, or anything that will change how this will work? And to be safe, you need it all written down.

You need to move bees, of course, and if you are expanding you'll need another way to move more bees, so take care of that first. Take more trips, get a bigger truck, or know somebody who has one. In any event, that cost has to be covered. If it's going to be more trips, include your time spent loading, driving,

Pollinating cranberries is a lot different than the local apple orchard.

unloading, and repeating in the calculation. Even if it's not far, it'll add up.

How do you know how much to charge? One way, though not the only way, to calculate this is to first determine the cost of keeping your colonies alive for a year. That's colony overhead. Add that 30 percent or so profit you need to stay in business, and that gives you a foundation value for a colony. Does your colony make honey on a pollination job? If not, would it have had you left it in a good beeyard during the same time? Are your pollination colonies as healthy when the job is done as those left at home? Do you spend extra to get them healthy again, or to get them ready for the down season? Those are opportunity costs that you have to consider, along with the time involved in simply doing that extra work. So you have, essentially, a cost your pollination fee has to cover, before profit. Plus, take into account the colonies you lose, because of the local bear population, for example, or pesticide damage. So you have a colony cost, plus the delivery cost to get them there and back, plus travel cost for inspections, plus your profit, and that's a ballpark for your pollination fee.

Got a contract, or just handshake, or even less? Many things can go wrong. Right off, what happens if the owner dies while your bees are in the orchard and there's a legal issue about who owns what? Then there are all the regularly occurring things: colonies run over by a careless employee, accidental sprays, stinging incidents, theft, and more. A contract should be a list of "what ifs," just so you're covered. Ask yourself if your lender would mind if your bees disappeared, if the owner didn't pay, and so on. Yes, many pollination deals are handshakes at best, and a written contract is never considered. But there are many sample contracts on the internet that handle all of these issues so take a look at them. Don't ignore a pollination contract, even if it's with your brother. Don't gamble your future on something your brother can't control, let alone a stranger.

And then there's the big show: perhaps almonds, northwest U.S. tree fruit, Maine blueberries, or Wisconsin cranberries. This is a whole different ball game. Contracts are almost always a given, and they are more than a page long. Before you begin, make as many contacts with as many other parties as you can: growers, brokers, other beekeepers, trucking firms, more beekeepers, and the like.

If you're still wet behind the ears, start by talking to brokers or beekeepers who have been doing this for decades. Often, other beekeepers act sort of like a broker when they need more colonies to fill their existing contracts. Yes, you can lose your shirt here. And yes, it can work wonderfully and you can make enough extra in March to buy that new truck. It's usually somewhere in between. But start with someone who's been there. Ask the hard questions: What went wrong? What would you do differently? Who can I trust? Who shouldn't I trust? Who else should I talk to?

If your colonies are being hauled to California, and they probably are, do you have some kind of arrangement with the people responsible for hauling them? Do you need one? Yes, you need one. If there's an accident your broken boxes will be replaced if you have the right insurance or agreement with the trucking company, but the bees that were killed by the emergency responders to fix the wreck will not. What happens then to your lost income from the job and, in fact, your pollination contract and your reputation down the road? And what about the honey crop they won't make this summer, the other pollination jobs you can't fulfill and won't make money from, the nucs you were going to get from them to sell and expand your operation, and the list goes on. Some trucking companies have all or at least some of this under control. Some beekeepers do. Lots don't and you are out

Heading to California to pollinate almonds is the big show. Know what to expect, and be prepared for the worst.

Who takes care of your hives when they are in an orchard? Do you have time? And who takes care of the store when you're in the orchard?

You have to move them with what you have . . .

. . . or what you'll need.

your bees and all the future income they would have made plus the money it will take to replace them. It's sort of like killing a pregnant cow; you lose the cow, the milk she would have produced during the rest of her life, the calves she would have had and their milk, and on and on. Back when you had a day job, you might have been able to absorb this and survive, but when it becomes a big, or even modest, percent of your income, it's a whole different story.

Beekeepers make more money pollinating crops than they do making honey. Have straps, entrance guards, and even nets available.

The same goes for their return. What guarantees or insurance do you have on that part of the trip? At the same time, even if they arrive safely do they arrive on time? If delayed, and the colonies are set for another job, who eats that loss? But what if they arrive right on time but the weather is still winterlike? If they have been arriving, or you expected them to arrive home and be ready to split, requeen, whatever, and you have to deal with snow, frozen ground, and no forage, can you afford the extra costs, the delays in splitting for future sales, and any lost pollination jobs because they aren't ready?

There are good points to all this, however. The flush of cash at the end of your cash season may be the best thing that can happen. Committing half your bees, half that you could actually afford to lose if worst came to worst, in trade for more than half the money you'll make in a year may be an investment you should make. What does all this mean? When it works, there will be lots of cash; all of your bees to split, sell, and harvest honey from; and a cash cushion for next year, when it doesn't work, there's no honey, and you end up with fewer bees. Look at the best and the worst way this could play out and if you can afford the worst, you'll have your answer.

Inspection is part of the game. Don't cheat.

Your colonies may get inspected for strength. Your customer will want every colony to have roughly eight frames of bees and brood. Before the process gets started, clarify the details: the time of

Power-wash colonies and pallets to remove any chance of weeds, seeds, and ants. Some even use new pallets for a move to California.

day of inspection, the cost, the outside temperature, who will do the inspecting, if there is any opportunity for reinspection, and who will do that. And what about extras? What if some small percentage of your colonies come in weak? Do you have extras to supplement them with? If not, what happens to your contract? One thing's for sure: you most likely won't be invited back. So consider extras. There's a good chance you will be able to use them with another grower, but even if not, you won't damage your reputation because you had a short load.

Then sharpen your pencil and figure out if this is worth doing. Will the cost of fixing what's left in April or May be more than you make in March? Will you lose more because the colonies just aren't in good shape, so you spend to fix them, and then just break even or even lose money the rest of the season? Could you find someone who will take some number of your colonies with his so you can measure the stress after they return?

What does your contract say about pesticide sprays? It happens.

Don't forget to go visit your bees. In fact, visit them long enough to see the holding yard, if there is one, and help move them into the orchard. Meet the grower. Take photos of your bees in the orchard with the grower or beekeeper, if you can. When something goes horribly wrong, it's evidence that will hold up in court. Consider the time spent an investment, because if they come home sick at least you'll have some idea why. If they don't come home at all, you'll know where to start. That happens.

You'll need clean colonies and pallets. Border inspections are getting more intense, and if there's an issue with weeds or ants and the truck has to be cleaned, it will have to be moved and cleaned away from the station. This could delay delivery and knock something off your payment. Some beekeepers are using either new or power-washed pallets in addition to washing the colonies themselves before loading the truck. That almost always guarantees a clean load,

but at what cost? You have that many pallets out back somewhere, right? And you have four people on hand to help get this ready for tomorrow, right?

The time of year all this happens is going to be the busiest time of the year for the rest of your business. Will you have staff to watch the store, answer the phone, help customers, check bees, and all the rest when you are out back getting things loaded or in you're California checking your bees? Maybe you're chasing another apple orchard in the same county you live in, but you have to be there tonight, so you have to get them ready today. And what about your honey colonies? Are they in need of feeding, medicating, moving, checking, or unwrapping? Who does that when you're in California, or on the phone with an irate customer who needs something you don't have?

What we're looking at here are the worst-case scenarios that happen with pollination timing.

These hives won't go home, but can be sold to other beekeepers for more pollination. This is a growing market.

Everything can go terribly wrong or absolutely brilliantly. Be prepared for the first; celebrate the last. But go in with both eyes open.

A commitment like this is going to change the way you do other things in this business. To have colonies strong enough at the right time of year is going to mean a solid mite-control program, which means treatment frequencies that get in the way of good honey production. So you either don't use as many colonies for pollination, leaving some behind getting ready for the honey flow, or you do what more beekeepers are seeing as an alternative choice: making bees. Not everybody is buying bees from some producer in the southeast part of the U.S. or northern California. Some are making their own now.

In the long run, what does happen to honey production for your business because of pollination? Which is most profitable? Which provides the least risk? Can you do both and make money in the end?

The commercial beekeeping industry is evolving and honey production at that level is vanishing, yet production at the local level is flourishing like never before. Can you better (or should you) concentrate on staying home rather than spreading out too thin to do any of these jobs profitably?

Moreover, staying home solves all the problems just listed. You don't need to hire extra people to keep the home fires burning, you don't challenge the health of your bees in holding yards, on trucks, or by disrupting seasonal cycles, and you don't miss locally produced honey crops. If pollination, or if ramping up your pollination business, is something you are considering, weigh all the factors, both good and bad, before committing to a life on the road.

Bees—Gotta Have Them, Gonna Want Them

There is a growing market for these additional bees. The ever-increasing beginner's market seems like it doesn't have a ceiling. Even if it's not expanding, it continues to be productive. Replacing 30 percent of all the bees in the U.S. every year has created a demand, not so much at the beginner's level, but further up the food chain. The bees used to make honey are being used more and more to make more bees. They're shaken for packages right after pollination, used for splits and divides all season long for filling holes, supplying drone source colonies, and beefing up for next pollination season.

This is happening because beekeepers with a lot of colonies who would normally produce honey aren't finding the barrel market very profitable. That's because, as you well know, less expensive imports are becoming extremely competitive in both price and availability. On the global market, honey is being seen as a cost effective way for landless people to make a living off the land, thus increasing their income and quality of life. Governments are encouraging this, and supporting the export markets that are a part of this. So the price of honey is down in almost every market, some more than others.

You may not run colonies full time, but I'll bet that for at least some part of the year, there will be bees in your life. After all, nucs, packages, whole hives, and queens still play center stage with your customers.

PRODUCING YOUR OWN BEES TO SELL

If you are in the position to reduce your personal bee load in the spring, early summer, and even into late summer, you have a distinct sales advantage: you know far more about the condition of the bees than you would if they arrived on a truck yesterday. Depending on your location, starting in early spring your job will be inspecting, splitting, and shaking your bees, readying them for honey production, making more splits, queen rearing, pollination, or simply sale.

Most states are going to require some sort of inspection before you start selling, so get that out of the way as early as you can. But don't do a one-and-done inspection. Bring the inspectors back, even if there's a cost, so you can say "inspected this month." Make a big deal about that; it shows you care.

What about selling whole hives? Why not? Some of your customers need to fill pollination contracts, make a honey crop, fill holes in their nuc production capability, or the like. First, define what you mean by a whole colony and stick to it. Is it one box, two, or three? How many frames of honey, brood, and bees?

Assembled, painted whole hives.

Overwintered nucs, strong and ready to go, are a goldmine.

Is it new or used equipment? What is the age of the queen? Once you can calculate what a full-size colony is, you'll know what you have invested in it, and you can set a price. You know your schedule, so set available dates accordingly. Plan around your schedule, not your customer's.

How old is the queen in that hive (or nuc) you plan to sell? This question challenges the ethics of more beekeepers than any other I know, and it really shouldn't. Here's a plan that will set back sales a bit due to timing, but will be effective in the long run: Keep last year's queens at home if they appear healthy and productive, and use them there. Do not dump them on unsuspecting customers who will lose them five or six weeks down the line and either blame you—as they should—or, more often, wonder what

they did wrong. Install a new queen and give her at least a month to get situated. Yes, a whole month. Another ethical way to do this is to install her late last summer and let her overwinter with all those bees. Transfer her into that high-quality unit next spring, laying constantly and ready to go. This solves the problem of having to install too early in the season and keeps a healthy bee in that box.

Whole hives are probably the easiest, earliest, and safest units to sell. But sell high quality and not just passable. First thing in the spring, transfer that overwintered hive from last year into new, painted wood, with last year's frames and foundation, just like your nucs. It'll be new wood and nearly new frames that aren't all gooped up. Decide on a standard strength you won't go below that might not be a buster, but is more than a five-frame nuc, add a couple frames of honey, and you have a profitable box of bees. It's not quite an eight-frame pollination unit, but it's not far from it. But what size to sell? If every brood box you own is ten-frames deep, the answer is simple, and unfortunate. So the first question is, what size are your customers mostly buying anymore? Mediums are popular and becoming more so, so if you have only deeps, you are going to lose some customers. Ten-framers are still more popular than eight-framers, but about 20 percent of beekeepers have at least some eight-framers. Don't be stubborn; sell what the customer wants. This means, of course, that you'll have to think ahead a season or two to get this going, so start this spring for next.

If you're selling a medium eight- or ten-framer, the customer walks out with a ready-to-go hive. You add in another two mediums for the rest of the brood box and maybe, down the line, a couple more for honey, and you're set. And so are they. And while you're at it, sell them feeders, feed, hive stands, tools, protective gear, new covers and bottoms, and everything else they will need to succeed. Give them a list. Have signs-up. Put displays all over the place showing how everything works and why it's better, faster, easier, and cheaper than what they have now. If customers don't hear and see, they don't buy.

 Don't forget gifts! Your customers' neighbors and, especially, family like thoughtful gifts that are also practical. A honey dipper or honey pot, a special kind of honey, or a hand cream. Honey sticks for the kids? All these add up to more sales for you. But you have to have signs so customers think *gift* and not *need*.

THREE-, FOUR-, AND FIVE-FRAME NUCS, MEDIUM OR DEEP

These are for the masses. This is the commodity part of selling your bees. How many you can make up in the fall, and again in the spring, is the only limitation to what you can sell. Nucs are popular and in demand, the most that they have been in the past thirty years. They're less of a gamble than a package, quicker to kick into gear for honey, and require no guessing, waiting, or hoping. Full speed ahead on these.

Mediums are hot, so you'd better have some ready. Deeps are still popular; which you'll sell the most of depends on your region. See if you can question someone nearby on this so you have a feel long before you start raising these. But don't ask what they sell; ask what the customers ask for.

SPRING, SUMMER, OR EVEN FALL NUCS, NEW BOXES, NEW QUEENS

Make splits from splits for a later colony. You probably do this already. The resulting colonies are too small for early pollination, and most likely not honey makers, but they can be late spring or summer nucs for sale. They are usually smaller, with three or four frames—about a third brood, a third honey, and a third bees and pollen. Like I said, small. But right away, you introduce a new queen and feed heavy protein and carbs with a stimulant like there's no tomorrow. They'll be ready for mid-summer sales to anybody who's planning to start expanding next season and wants a head start now. Make sure the new queen is producing at least 500 eggs a day when you hand her over. If it's less, wait another couple weeks. When you make that split, keep the box from the mother and put the new colony in a new box. It looks better and makes it more valuable.

Honey, pollen, open and sealed brood—all necessary for a good nuc.

BROOD AND QUEENLESS NUCS

These are possible products, too. Some customers need brood for their own nucs because they came up short or are boosting colonies for pollination. Pricing a frame of brood is easy. A deep frame consists of about 9,000 cells. A three-pound (1.4 kg) package has about 10,000 bees that are about three times as old as the bees in the frame. So the frame is priced at a third the cost of the package, if it's absolutely wall-to-wall brood, minus 10 percent because there are 10 percent fewer bees. If it isn't wall-to-wall, what percentage is missing? Knock that amount off your full price, and you have your actual price. For a queenless nuc, charge simply the cost of a nuc, minus the cost of the queen you would have put in it, whether it was yours or from someplace else. It's easy, and there's no need for guarantees or returns for either.

TRANSPORTATION

Most of your nuc customers will be prepared to get a nuc home—probably not on the front seat of the car. But what do you do with those who didn't figure this out before they left home? My guess is that they pre-ordered that nuc via phone, email, or a personal visit. That was your opportunity to give them the heads-up that they needed to make the best choices for moving that nuc. In fact, that was your opportunity to tell them a whole bunch of things about nuc care and feeding. Get an email address when you take the order, and send each customer a prepared email with a booklet on the care and feeding of that very expensive toy they just bought. They may never read it, but they may buy everything you list, too, if they didn't know they needed it. On nuc day, you simply won't have the time to discuss all this with them.

SAMPLE NUC SELLER'S HANDOUT

Here's a bullet point handout (or email) you can give to everybody that buys a nuc.
(Make sure this is customized to your location and seasonal schedule.)

- Prepare the location where the nuc will sit before you leave home to get it

- Travel home with care. Secure the nuc and make sure there is more-than-adequate ventilation so the bees don't overheat

- Even if there's honey and pollen, feed, feed, feed—both carbohydrates and protein

- After a few days, have your ten- or eight-frame equipment ready, with drawn comb, if possible, or foundation if not. If plastic, brush with melted beeswax to move the bees along

- Transfer the frames and continue to feed

- Continue feeding until foundation is drawn and honey is being stored in new frames.

- Monitor the queen, and when there are four frames of mostly sealed brood, divide this colony in two, with equal amounts of brood in each

- Have a queen ordered and requeen the queenless split

- Feed, feed, feed both

- When each split has at least four frames of sealed brood, divide both again. It should be about mid-summer if you started early enough.

- Either let each raise their own queens, or requeen when a sealed queen cell is detected, leaving the colony broodless for that long

- Before requeening, test for mites. If more than one mite in 100 bees is detected, apply your treatment of choice.

- Requeen and feed, feed, feed, feed, especially high-quality pollen substitute of better, fresh pollen just collected

- You should now have a beeyard full of colonies all from a single nuc

A truck is always a first choice for getting that nuc home, of course. The nuc should be faced so the frames run front to back instead of side to side. Secure it to the side or back with a hold bar so it can't move during starts and stops. Use a cinch strap to hold the cover, box, and bottom to keep them from coming apart. Place a mesh sack over the whole thing so it doesn't overheat but keeps bees in. Wedge a screen securely in the front door and hold it in place with a pushpin or nail. It will have good ventilation, it will be safe from moving around the bed of the truck, the bees will be securely inside it, and everything will hold firmly together. That's the way to go. I'll bet you skip some of those steps sometimes, don't you?

Make up a unit like that, take a picture, and put it on the set of instructions you send out when that customer orders that nuc.

But then, we don't all have trucks, do we?

A nuc secured with a cinch strap and a stable screen in the front door will almost always be okay fastened to the back seat of a car. Never, ever put it on the floor. Driving with the windows open and air conditioning blasting will be just fine. A screened top is better in the car, and should be considered if traveling a long distance. Do you have some of those to sell?

That booklet you sent via email should have moving instructions in it, and you should have all the equipment you talk about, ready to use. Sell inexpensive cinch straps; the screen entrance and pin are on you. An accident because of poor judgement on your customer's part is a terrible thing to have happen. Don't let it.

FRAME EXCHANGE

If you haven't already been doing frame exchange, don't start now. You only want to sell year-old frames from honey supers, and you don't want anybody's trash in exchange. You're only buying poison, disease, and problems. Get rid of your older brood frames on a three- to four-year basis, newer honey frames in your nucs every year, and keep everything as fresh as you can. Don't even think of frame exchange.

QUEENS

Raising queens may be a headache you don't have to have. Selling them is a headache you have to have. If they're your queens, the headache can be much, much worse, or much, much better. It's complicated.

If you haven't raised more queens than you need for your operation, this isn't the time to learn how That's for another book and class. But if you are selling some you raised, keep the cost basics in mind. Time and equipment are certainly necessary, especially when it comes to mating nuc management and producing very strong drone colonies.

But selling the queens you produce requires the same set of ethics that selling any living animal requires. That queen has to be perfectly healthy, has to have been laying for at least a month, and has to have a perfect pattern. Her first flush of offspring needs to be as healthy as she is. She, along with the colony she is raised in, needs to be absolutely free of anything bad, and you have to be able to show your records to demonstrate you are on top of things. That once-a-month visit by the inspector you insist on really pays off here.

You also should have records of the tests you have had performed on your operation—for instance the tests done at the North Carolina Bee Lab on the queens from her flush for size, sperm count, virus load, and the like. If you can't show me those, my first instinct will be that you have something to hide. Maybe a lot to hide. I want to see Beltsville Bee Lab reports on how clean your bees are of nosema, Varroa, AFB, and viruses. I want to know this queen,

TIP: Two tests that should be routinely run on queens are a hygienic behavior test and an egg production-per-day calculation. To test for hygienic behavior, take a file card and cut a diagonal square that shows exactly 100 cells (there are test kits that have this already made). Hold the card over a section of sealed brood and kill, using a pin, all the brood in that opening. In five days return and count the number of larvae that have been removed. Repeat in ten days. Compute the percent removed each time. Aim for greater than 75 percent the first time and nearly all at the end. Egg production is easy to measure. First, count; count every cell of sealed brood on every frame. Using your cell phone, take a photo of every frame and count them later. Exactly twelve days later—at the same time of day—do it again. Subtract the second number from the first and divide that final number by 12. The result is eggs per day.

Mating yards mean work. . .

this very expensive queen, is worth the money. If you don't have, can't get, or won't get that data, don't sell queens.

FINDING A PRODUCER

If you're not producing your own bees, you'll be getting hundreds, maybe thousands, of queens through commercial queen producers. All of them will be, or should be, from a queen producer who sells queens with a perfect record of experience, background, and reputation. If they can't, find one who will, because that queen has your name on it. You thought highly enough of that producer to buy and resell those queens. If you aren't smart enough to find a good producer, why would I trust you to make any good decisions on the rest of your stuff? So start with the best—not the cheapest, not the earliest, not the fastest—the best.

So, who's good enough? Are you going to put all your eggs in one basket? Maybe your package pro-ducer is. That could work for a lot of reasons. Maybe someone you've dealt with forever and trust? Good. Someone close? Maybe not. Find someone who has a reputation for reliable stock, is easy to work with regarding delivery time and returns, and of course values quality and service. Ideally, work with a couple of suppliers who meet those criteria, because stuff happens out West one year, down South the next.

And as with honey, don't sell generic queens. You should have a west coast variety, a southern variety, and if it works, a local stock. They should be advertised and priced differently. "My Western stock is tested and used by more commercial beekeepers than any other queens available." "Our Southern queens are raised in the heart of Historic Queen Country by seventh-generation producers." Give them an identity. Be sure to have some experience with each, so you can honestly tell people what to expect. You can make them available seasonally: Southern early, Western a bit later, and local mid-season.

. . . and a lot of time. If you run a store, do you have this kind of time?

Local queens can be tough for most sellers because of availability and timing, but if you work it right, you can use that to your advantage by selling them as package replacement queens or summer split queens. Make it known when you'll most likely have them, so your customers know what to expect. If local isn't what they want, then a holding split is what they need when you have early queens for their later splits and replacements. Stress that management technique so they have them when they need them and you're not scrambling in July.

Find a back-up producer who can probably help out with small orders later in the season. Prices for these have to be higher because of your customers' real or imagined panic, but because of who you know, you can take good care of them.

STORAGE

If your queens are going fast, you can leave them in the battery box, making sure there's more than enough food, the temperature is right, and you're adding fresh broodnest nurse bees daily. You can't just add foragers; they have forgotten how to make queen food. You *must* have broodnest nurse bees. But what qualifies as moving fast enough? With fresh bees and enough food, they should be fine for a week. Much longer and you'll start to see some problems.

Another option is a queen bank—a queenless colony with lots and lots of young bees, just like your producer uses or you use in your operation. In fact, it's much more impressive when a customer wants a queen and you take them out to that colony, open it up, show them the cages, and ask which one they want. (Yes, give them a veil, but 999 times out of 1,000 you won't need one, so don't you use one.) Let them pick it up and take it home. They absolutely own that bug when you do that. And you can bet they'll do everything you tell them to do to get her accepted and keep her healthy.

Requeen your nucs at least a month before the sale.

The same can be said for selling cells. These won't go to the second-year hobby beekeeper, but you probably have some longtime sideliners who know more than enough to get a cell from your colony to a mated queen in theirs. You know the timing, so you've already sold this bug days, weeks, maybe even months ago. Well, money hasn't actually changed hands, but the buyer is expecting those cells during a particular week in April or May and is waiting for the call. When it's time, you call, they come over, you hand over individual cells, or maybe a bar with several on it, they put it in a safe and stable and warm place, and they're gone. Assume 80 percent of them will get mated and be laying in 10 days; that's what you are selling. Up front, tell them 20 percent probably won't make it, and sell them at that rate. How much? Half of what you'd sell a queen for. And then 20 percent off that, with no guarantees at all—out the door and she's all yours.

Packages are shaken on a Friday . . .

. . . and loaded for the trip in environmentally controlled trucks.

When they arrive, they are fed and customers are notified . . .

. . . and Package Day happens!

Do you mark your queens? For free? For a fee?

PRICING QUEENS

You know how much you have to have to pay for a queen, your time, and your losses. Get out your pencil and paper. And no, you don't have to care what anybody else is selling them for. But here's a question: do you discount a bug you've had too long? The first and only answer is no. Never. Make a two- or three-frame split and keep her yourself. Don't compromise your reputation and the quality of your products with an inferior queen. Never.

RETURNS

A lot of bee sellers use a taillight guarantee for a queen. Once she's out the door and I can't see your taillights, she's yours. There are no returns for any reason. The variables that can lead to the death of a queen are legion, and inexperience is most often, though not always, the cause. Here's the deal: If you know your producer and know that what you are getting is the best you can get, then you are selling the best there is. If the bees have been tested and you are confident in their quality, then you can reasonably stand by a no-returns policy.

Until you can't. Suddenly you may find a lot of problems. Maybe your producer simply had a bad batch. Maybe there was poor mating weather or food shortage. Maybe he got his pull time from the mating nucs screwed up, the bees got overheated in transit, they froze on your loading dock, or one of your kids wasn't looking and dropped the battery box. A million things can and will go wrong. So then what?

Make sure you've discussed this with your supplier beforehand. Ask the "what if" questions before you need to know so you know what to do when it happens. And get data. Know everything about your customers, where all those queens went, how long you had them, how long before you heard from a customer—everything you can think of so you can support a recall or refund for that batch.

Just like your operation, the season ahead rests on the quality of that queen. Some of your customers are going to have a less-than-stellar season because their colonies just never took off, or simply died before they got off the blocks. For the hobby folks, a bad queen is just one more avoidable obstacle to overcome. If it's the third year in a row, you can bet they won't be back next season. You've lost a customer, and that customer—remember word of mouth—will lose you some more.

PACKAGES

Selling packages is almost as old as keeping bees. And it's not much more complicated. Let's take a look.

You've bought a thousand packages over the years. Most were okay, some were bad right off, some crashed later, most but not all of the queens did okay, and it was the fastest, easiest, cheapest way to get more bees. Like other people's queens, other people's bees are changing. Bees used to be a commodity. Producers in the South and the West had a jump on the season and could grow bees like grass, and for a long time that was taken for granted.

But that's changed, in some places more than others. African honeybees changed it first. And *Varroa* and its viruses have changed it more. The chemicals used to control *Varroa* have changed it again, and issues with nutrition have had a say in the matter, too. The bees that are in our packages aren't the same as they used to be. So know that going in.

Where will you get your packages? If you've been hauling your own for a while and know your distributor, the decision may be simple, if your distributor has as many as you are probably going to need. If not, what then? Some here, some there? All from another producer? You need to sort that out probably 2 years, maybe 3, ahead of time. It's easiest to stick with your first producer. Stay at the same amount the first year, and tell them you will increase your order next season and by how much, so they can fit you in. It's when they top out that you have to decide where to go.

Having two suppliers isn't a bad idea. Different kind of bees and queens can be a promotional advantage to work with, and two delivery dates may reduce the package day circus a bit, too. If you are familiar with both suppliers and can get enough the first year to make it worth the cost, give it a try. Maybe try a couple of times, to see how the second supplier

Have ways to move a bunch of packages, over all kinds of surfaces, at the same time . . .

. . . and a safe place to put them.

FIELD DAY IDEAS

Depending on where you are, if possible, hold a field day to install a package before the bulk of your packages arrive. You can make up a package from your bees for this. Show everybody how to do it the way it should be done, using the kind of package (wood or plastic) participants will be getting. If you can't do that, hold a field day shortly after the packages arrive and show after-installation care, especially how to check a queen still in her cage, and, especially, how to install a new queen if the first one is lost. If you can, just before the class, replace an existing queen that has been accepted, with a new one so when the class begins the bees in the hive will be acting aggressively toward her, so you can easily show that behavior, too. Have several feeders available and several feeds to use, and do a smoker lighting demonstration, again, just so they see it, again. And make sure everybody gets to hold a frame, smell the bees, and hear the hummm. It will be a day they won't forget, and that will go a long way to ensuring they do things the way they should be done so they succeed.

works out. You might be working with them all the time if your first producer doesn't expand, if having two works better, or if they are better bees.

Depending on volume, a third supplier might even be an advantage. They'll have different dates, bees, prices, and queens; and giving your customers a choice is always better than not. Far too often, we get the bees we can, rather than the bees we want.

You have to be prepared for a crash and burn from one or two of your suppliers on occasion. Crop failure isn't restricted to just cotton and corn. Having more than one is looking better all the time.

What kind of guarantee are you going to offer on packages? Bees die, queens aren't accepted, packages abscond, and stuff happens every year to somebody. Some of your new customers are dumber than bricks and will kill a colony dead in a week. Then what? Maybe just the queen dies in her cage in the first week; then what? There's always the taillight guarantee (see page 120).

The plastic Bee Bus is the future.

Here's another option: Your local association probably offers a beginner's class every spring. The quality of the teachers is all over the map, with egos and dollars and experience often getting in the way of good education. Make those beginner students an offer. After the beginner's class, if they take your two Saturday morning classes (or maybe your longer single Sunday class), you will absolutely guarantee the packages they buy from you. Make tuition the price of two queens, and you'll come out ahead every year. It requires only two Saturday mornings in February with a bunch of I-can't-wait-to-get-started beekeepers in your store looking, touching, trying on, asking, and most importantly, learning how to expertly install and care for those new packages coming in May, which they ordered from you in January as part of their local association class. If the package absconds, total replacement is free. If just the queen fails, you'll replace her for half price. Make that offer to every first-year beekeeper who buys a package from you. Call it your "Guarantee with Class" so it's easy to remember.

PACKAGE DAY

Package day may only be one day, but the planning starts months in advance. Start promoting it when your catalogue comes out. Do mailings, emails, and social media announcements all year long. Give it as much attention as you can. This is the center stake in your year. It will certainly make (but probably not break) your season. Promote.

Touch base with every customer who ordered a package a week or so before the big day. Send an email (you got that when they ordered) earlier so you know it works. If they don't have email, phone them. Touch base so you'll know you can get ahold of them just before the packages arrive. When you get confirmation that the packages will be at your place and ready to distribute the day after tomorrow, let everybody know exactly when they should be there, what they should bring to get the bees home safely, how much they will owe you and how to pay, and what they should do when they get home that day,

the next week, and the next month. (You put this together long ago and sent it as a PDF attachment in the email; see page 114. Nothing more on this … just do it.) One more light in the taillight guarantee.

Just to be safe, have extra queens available before package day, maybe from the package supplier or another source. There will be some packages with questionable or dead queens, and you'll have immediate replacements on hand.

Then, be ready. Work with your supplier, or work with the people delivering the packages (yes, it might be you doing the driving, but keep in touch back home) so you know the time of the day they will pull into your driveway. Your crew should be ready. Your supplies should ready. You have somewhere out of the sun, maybe even air-conditioned, to put all these bees, right? Have hoses at the ready to keep those bees cool, and maybe even a very big fan to keep a breeze on them between soakings. Cool is the name of the game with these pallets of bees. Keep them fed, too, especially if they come in and you have to do some separations beforehand. Italians here, Carnys over there, Russians way over there, queenless packages somewhere else, and those we mail tomorrow close to the front.

And you have people to help, right? You'll need lots of people: two people at the checkout, because not everybody is buying only packages; someone outside or wherever the bees are, making sure nobody is messing with them; runners to get the packages for every customer; wagon drivers to move the largest orders; people marking queens; somebody checking every queen to make sure it is alive (remember that taillight?); somebody helping load cars and trucks; somebody helping customers find the stuff they want; and you, making sure it all goes smoothly. You've been to a hundred of these, but only as a customer, getting in and out as quickly as possible. Now, you have to make sure your customers are getting in and out as quickly as possible. It's you keeping the gears greased, the crowd moving, the packages taken care of, and the customer cars into and out of the parking lot.

Take as long as it takes to get this day finished for everybody. Schedule arrivals so about half the customers show up in the morning and the other half comes in the afternoon. It helps. That night, when it's all over, the last customer has left, the remaining bees have been cared for, and the place has been put back together with the floor swept and the doors locked, treat your helpers to a huge, terrific barbecue. Let this be a wonderful day for all of them. They worked hard for you, they helped a lot of beekeepers get going or keep going, and they deserve it. And they'll come back next year and do it again.

 During the long, hectic, and crazy package day, be sure to rotate your employees for all the jobs that need doing. Don't stay at the cash register all day, don't be marking queens all day, don't stand all day and hose bees, help load trucks, show how to keep the packages safe in a car, or refill shelves. A change of pace every couple of hours goes a long way in making a hard day easier.

BUYING NUCS

The demand for nucs has been increasing every year because beginner class teachers keep telling students that a nuc is better than a package. With demand going up, the number of people selling nucs is increasing. So let's talk about buying nucs from somebody else to resell. First thought: Don't. Unless it's from your brother and you trust him, there are just too many variables with this. You need to triple-check every detail to make sure your customers—not his customers, *yours*—are getting their money's worth.

Do you check every nuc for a good queen? How can you tell if it's a good queen? Was she released yesterday afternoon, and the eggs and open brood are from some other colony? Is there disease? Comb loaded with chemicals? There are a ton of other things that can, and will, go wrong. And all of these will cause the nuc to fail, or at least not thrive, like everybody said it should. And your customers will be back to ask you about this while your producer is on his way to the bank.

But if you insist.

If you've been working with someone for years and it's been good, don't change. But if this is new, here are some thoughts: First, start this process the year before with clear, easy-to-interpret, and easy-to-implement guidelines for the producer that you both agree on. There should be standards of strength relative to bees, eggs, open and sealed brood, age and type of queen, comb condition, age, and maybe others that are easy to meet and easy to measure. You might even have different categories to spell out that will sell at different prices.

For instance: There is a difference between overwintered nucs with year-old queens and splits from this season with a queen just stuck in—and where is that queen from, by the way? Will you get old comb, second-year comb, or foundation? Wooden boxes to return or paper to keep? Three frames with bees or two or five? Frames of brood, frames of honey, empty comb, no comb . . . how many of each goes in a five framer? What about frame exchange, and queen loss the first two weeks? And is there any guarantee at all on any of this? Have the nucs been inspected by someone other than you and the producer? Do they have to be?

This photo says it all about parking space. Do you have that space available, or close enough, to borrow for a day? And do you have even more parking for the customers? And do you have enough help to take care of the nucs, get them loaded, answer questions, and make sure they get paid for? Be prepared.

If you're just starting out in this business, even if you've been working with bees a long time, selling somebody else's bees during the first couple of years can be tricky. It's a lot of extra work that will take time from your other chores. That said, it is definitely doable, and if timed right, it's both profitable and a great service to your customers, especially if you have more of a bees store than a supply store. Those whose business happens mostly during a couple of months in the spring as they make, get, and sell bees should jump at this, because that's what they're known for—bees. But for those who are trying to figure out all the ins and outs of running a store, hired help, packages, woodenware, and all the rest, other nucs can be a real challenge.

This isn't to say that all nuc producers are less than satisfactory. Not by any stretch. Lots are excellent beekeepers, or they wouldn't have the bees to make nucs, they wouldn't overwinter nucs, and they wouldn't raise their own queens for those nucs. And fortunately, there seems to be more of them every year.

The back office. Make sure you have one.

To determine the big picture for your operation, and this has to come first, consider how much you can take on, initially, how much you want to take on, and what would happen to your business if it all falls apart? You really don't what to fail because someone else didn't do what they were supposed to do.

BUYING OTHER QUEENS

If you sell queens you raise yourself, then you already understand how this works and probably have a replacement policy your customers understand. If not, you need to develop one. The taillight guarantee works here for a lot of people. But remember that class you taught earlier this spring? You can make that work here, too. If a buyer has taken your package class this year, when just the queen in that package goes AWOL, you can get them a replacement for half price. You'll eat some queens, but not many (after all, they took your class!), and it makes your business a responsible member of the beekeeping community.

You'll most likely bring in battery boxes of queens from several suppliers. You'll need some level of good environmental storage. It should be cool, but not cold, and shady with no direct sun. It should have good air circulation, routine feeding, dead bee removal, and all the rest of good biological practices used with queen handling. Queens for package replacement and for requeening should be available as early as possible for those who want to make early splits, and as long into the season as possible for those late summer splits for *Varroa* control. (Do your part to promote that as a way to deal with *Varroa* without using chemicals in the hive.

ORDERING PACKAGES, NUCS, AND QUEENS.

Make sure this cog in the system is foolproof. Will you take orders on your webpage? By phone? Mail? In person only? Make those decisions before you begin, so the very first order—and every one after that—works perfectly. And then do dry runs on all of them. Develop a centralized system so no order gets lost from any source. Do you have a prepay policy? Should you? Work with your producer to make sure everybody understands the process of who gets paid and when. When you are sure it is foolproof, make it known that you can take orders.

How much should you charge? You have to know that several months ahead of time and for nucs, maybe a year ahead of time. Your supplier may have next year's prices already, so you can start there. Is that price guaranteed? If so, you can begin estimating volume by presales, walk-in sales, what next spring's local beginner's class size will be, and how many people sign up for your Guarantee with Class. Look at what last year's prices were, what the competition charged, what other sales outlets charged, and how much markup you absolutely have to have to make that profit per hour you need to keep the lights on.

Then consider this: How much would you sell one package for? Just one. You may think it's a ridiculous exercise, but do it anyway. How much for one mailed to you, one you drive to get, or one hand-delivered by the producer? You know your cost in product, labor, and storage. Start there.

Early on, determine how many packages your supplier can deliver in a load, and how many deliveries can be made before it's too late (and too late depends on your weather, your supplier's weather and luck). Once you have that number, you can begin taking orders to fill the delivery number. Here's a trick: Always order some number of extra packages for yourself. You will have last-minute customers, losses, and mistakes made in taking orders, and this cushion will cover you with your customers. If you end up with extras, you can make nucs out of them and sell them later in the season. Given typical care and feeding time and cost, you'll break even on them,

> **TIP:** Get the price of your packages, queens and nucs, and a close approximate to due date in your catalog. But encourage buyers to give you their email address or phone number so you can immediately notify them of changes in schedules. No excuses.

but you will have replacement colonies, extras for those who want to expand, and maybe a field day special come early fall. No matter what, you will have filled all the orders, and extras, and supplied nucs for later. It's a win-win for your business.

You can actually plan on this and make it known early on that you will have nucs for sale later in the season, set a price in your catalogue, and take orders the same time you take package orders. An advantage with this plan is that you can special-order queens your normal supplier doesn't or won't have and you can bump up the price a bit. Use the package queen for replacements for lost package queens from your customers, and put these specials in the nucs you make later. This works very well with "local" queens, that normally aren't available until a bit later than the packages. So, you can take advantage of extra bees, and special queens, and support the local queen industry and the local drone population all at the same time.

Bees—gotta have 'em, and you're gonna want 'em. But don't let them eat you out of house and home because of bad preparation, poor product, inaccurate pricing, and/or wishy-washy replacement guarantees. Do your homework before day one. Bees are why you're in business; make sure they aren't why you go out of business.

4 Promotion

How much business can you afford? That's the question you have to answer when you look at your advertising and promotion plans and budget. Those plans and that budget should cover everything you do to attract attention and to introduce new and existing customers to you, your business, and your services.

To run your business, you have to offer benefits: products and services available for people to purchase that are of the quality and prices they want so they return for more. We've looked at most of the products and services you could offer, at least eventually, and we've danced around products of differing qualities, and discussed prices that you have to have to continue. What we haven't looked at is where and who are all those customers are, and why should they purchase your stuff. And why would they come back?

Finding Customers

Customers first have to know you exist, and then you have to give them a reason to come back. Being the only game in town used to be a distinct advantage for a lot of businesses, but Amazon and every beekeeping business in the universe will ship you exactly what you want in three days or less, and at a price that's competitive with everything you have in the store. So no matter what words you use, your business has to let existing and potential customers know what products and services you have, where you are, when you are open, and what you can do for them. "Build it and they will come" is a pipe dream.

So, where to start?

Here are some basic questions you need to answer for the retail business you will be promoting. This will point you toward how your advertising and promotion activity will be most successful.

- The fundamental question for any business is, who is your customer? What do your customers want from you? Your Business Plan (see page 13) will provide much of this information; perhaps it's beekeeping equipment, bees, queens, nucs, honey, all the lotions and potions you make or

Is this what your customers expect?

bring in, and/or artsy stuff for candles and other wax products. Note that you will have customers from several backgrounds, and you'll need to meet their very different needs.

- Are there several, a few, or no resources your customers can take advantage of for further education on beekeeping, or on any of the crafts you support?

- What time frame will your customers shop in? Are they mostly weekenders or would evening hours serve them better? Is there a day midweek you could close and nobody would notice?

- Is your immediate neighborhood likely to change, affecting what you can and can't do in terms of parking, loading, foot traffic, and the like?

Educational signs are good for customers.

- If there is another beekeeping business in the region, does yours sell significantly different products and services? Are your prices, quality, selections, and hours so alike that you are cookie cutters, or are there major differences, such as a honey and bee focus as opposed to a wood and equipment focus? Basically, could you put your competitor's name on your ad and have no one notice?

- Do you have a suggestion box? Why not?

- You most likely belong to several local, regional, and national associations. Do you help administer some of them, along with reading trade journals to keep up with what the world is doing out there, so you are prepared both to answer customer questions and have the newest gadgets available?

- Do you look at every beekeeping equipment supply catalogue you can find, every year?

What Are You Selling?

Another set of questions to ask yourself has to do with where you get what you sell, and its perceived quality. We've discussed woodenware and bees coming from several places, but what about all the rest? Source will, in large part, determine what you have, what you can sell it for, and the quality. Do your customers know the source of your products? Worse-case scenario, they know it because it has a less-than-stellar reputation for being mostly cheap or mostly nonfunctional.

- Will everything come from the supplier you are a dealer for?

- Do your customers recognize your suppliers as being the best in the business or are they anonymous to almost everybody?

- Do your records give you a feel for how well all of your products are selling, all year long?

- Do you have to discount your prices to compete with the internet because your customers are only price driven?

- What's your policy for offering discounts for volume purchases? Or for local groups?

- What are the most and least price-sensitive products you sell?

- Are your customers often, sometimes, or never impulse buyers?

- Can you list all the services you offer, those nontangible things that you do?

- What's your return policy? What is it for equipment, bees, and lotions and potions? All the same?

Make sure the products and services you offer distinguish your business from other bee businesses in the area.

Customers need to recognize your suppliers as the best in the business.

- Answering these questions gives you a pretty good list of the benefits you have to offer your customers in terms of products, services, and education. Now, before you get too carried away, put yourself in your customer's place. You've been a beekeeper for years, and you know what you want and need in terms of service, products, price, availability, advice, and open hours. Are you supplying all of that? Is this a business you would go to and, even better, recommend to others? When you are satisfied, it's time to let the world know what's going on.

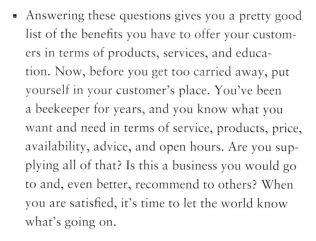 Strive to be exclusive in something; the "only one" for something. The only one in the state, the only one in the Southeastern United States, or the only one that sells all the parts—the only one. And then push that in your catalogue, in-store signs, and social-media bursts. Be the only one for something.

Advertising

You've heard the adages: "Advertising doesn't cost; it pays" and "What happens when you don't advertise? Nothing." They're correct. It's time to get started advertising.

THE PRESS RELEASE

A good promotional program begins several months before you open the doors to your store. Start with a press release telling the world that you are here to save the world of honeybees and beekeepers. When your operation opens—and you do need that business name now—you will be providing a whole list of benefits. Your release will tell why, now that you are here, bees, beekeepers, honey lovers, crafters, and others will be better off because now they can save money, make money, make more honey, provide employment when they expand, increase the yields of everybody's apple trees because there will be more bees, learn new skills, and meet new people. This isn't the time to hide your business under a bushel, as the saying goes.

Send another press release as your plans fall into place and you have a better idea of what it is you are going to do and who you will be working with. Keep your business in front of people, but emphasize a different set of benefits each time. Those benefits could be, for example, bees, beekeeping equipment, beekeeping education, lotions and potions, candles and ornaments, honeys, gift baskets, crafting classes, good store hours, health foods, and social events.

This can be an easy and inexpensive endeavor, or it can cost you some money. Either way, you have to do it. You've defined your customers pretty well by now, so find the media outlets that they need so you can effectively reach those most likely to benefit from knowing about you. It may simply be the local weekly newspaper, or it could be any of several media: weekly, daily, and shopper newspapers; radio and TV stations; local and trade magazines; and certainly any beekeeping newsletters that are around. We'll talk about social media later (see page 140), but don't forget that avenue, either.

Once you have a list of media to target, figure out the person responsible for getting press releases published at each one. Small, hometown weekly papers usually have small staffs, and you can go right to the publisher or editor. Larger dailies have departments to work with; you'll probably go to the city editor at a paper, and the news director at a radio or TV station. For trade publications, beekeeping magazines, and farm publications, you'll go to the managing editors. You probably know the local bee club's newsletter editor, and you can send it directly to that official. Find out who these people are and send the release directly to them. And keep this in mind all the time: you're helping them look good by letting them provide their readers with this wonderful bit of news.

Start with a cover letter. It should be short, a half page at most, but it will tell the story of why this release is important, and why it will be of interest and helpful to readers/listeners. Honeybees, and all pollinators, have been having an incredibly difficult time lately, and now, finally, there is going to be a local business set up to make life easier for them and the good people who care for them. Then add in a bit of your history and why you are the best person to do this. You want to do this because you, and now the newspaper, see there is a need in the area; because of what you can do, there will soon be a way to address that need.

Start at the top of the page with, "PRESS RELEASE." Under that, put the date—not the day you wrote it, but the day the news director will get it. Under that, state in one line what this release is about. Don't make the director have to work to find out. Make it easy. So say something like: "NEW BEEKEEPING SUPPORT BUSINESS TO OPEN IN MAY. HELP ON THE WAY FOR MEDINA'S HONEYBEES." You are selling the benefits your business will provide for the readers of that paper. Nothing else.

ADDITIONAL ADVERTISING OPPORTUNITIES

There are lots of things you can sell that beekeepers want and need, but you can't do it all, so take advantage of that and work with local businesses to cross-promote your products. Examples abound.

- You can probably sell the hive stands that sit on the ground and are mostly a waste of money for new beekeepers. But what about promoting that used building materials outfit just out of town that will sell used 4" × 4"s, railroad ties, and the like for pennies on the dollar to make DIY hive stands. You don't have the room for inventory, so let them sell them. Make a hive stand out of their materials and put it where customers can see it. Put a sign on it telling them where the wood came from. And do the same at their place: build a hive stand, add an empty hive with your business' name on the hive. It gives beekeepers good ideas and helps both businesses.

- Try something similar with all the garden centers in town. They sell fruit trees in the spring that need honey bees for pollination, and you sell honey bees. Put an empty hive at the garden center very near where the fruit trees are and put a sign on it that asks if there are enough bees in your yard to make this work. At your place, either set up a few still in pots when they are in bloom or actually plant a few different trees, with real hives right next to them, add signage that includes the variety of the tree and the name of the garden center.

- In midsummer, when your pollinator garden is blooming, put a hive close by and put a sign in the garden identifying the local source of the seeds. And do the same at the store that has its own garden and a hive of yours with your name on it. You can do the same inside the store with your honey, and add a menu from a local restaurant that uses your honey in its recipes. And, on that menu is the kind of honey used in that recipe (which is why it is so good), and where it came from.

- Work with your local paint store and promote their off-color paints that have been returned by customers that they will sell to beekeepers cheap because they can't sell it to anybody else. Have a multi colored stack of boxes with a paint can on top, display the name of the store where the paint came from. And, of course, they can do the same in their store. They will sell more of that paint, and your customers will be reminded of where to get cheap paint.

- Finally, you can do the same thing with Styrofoam or other types of insulation when it comes time to wrap hives in the fall. They get your customers' attention about where to get the stuff, and you attract their customers' eyeballs at their store.

Keep your first press release fairly general. State approximately when your business will be opening, where you will be located, what kinds of supplies you'll carry, and that you are working with members of the local and not-so-local bee clubs to help everybody do a better job of taking care of the bees and providing an exciting new resource for crafters and honey lovers. Everybody—homeowners and their gardens, wildlife, wildflowers, farmers and their crops, and people who like good food—will be a little better off because you are opening this new business.

Make it letter perfect—no typos, nothing to distract from the message. In the body of the copy and at the bottom of the cover letter, include your name, address, phone number, website, and email address. All of them. Make it simple to find you when they want more information. But be aware that those who manage the publication may not care and not publish it.

Mention the basics in your press release: "Survival Queens Will Soon Be Here!"

If it is published, send the person you sent it to a thank-you letter, again letter perfect. If your release is not published, or if the editors did their job and edited it, do not complain or call to find out why. If it wasn't published, try again after a short while. If you happened to hit a busier-than-usual week, it may have gotten lost in the shuffle. A second attempt within a couple of weeks may have a better chance, so time your fist one to accommodate this delay. If that too fails, it's time to invest in an ad in that paper that does essentially the same thing as your release. You may have the skills and software neces-

sary to put an ad together, or you may have to hire someone to do it. But do it. Add a bit of art—bees, butterflies, flowers, and beehives, maybe a beekeeper in the right gear in the background—and say essentially the same things. Offer a benefit to the readers, and let the local beekeepers know you will soon be available. Often (though not always) if you pay for your first announcement, you'll get a bit more attention for the second.

When you have more specifics, send out another release with more detail. You'll know who your suppliers will be, the hours you'll be open, the range of goods you'll have, and how many people will be working with you. Get quotes from your suppliers, such as "Glad to have them on board to help us help beekeepers even more" and from beekeepers happy to have somewhere they can go in emergencies and for supplies, not to mention a friendly face they can work with to make bees and beekeeping even better for everybody in Medina County. Again, offer only benefits to the readers. Why is it good that you are opening this business? Who will be better off, and what problems are you helping to solve? This one might not get published either; buy ad space again if it doesn't so the word gets out.

Make your news release cover letter look as important as it is. Use a #10 window envelope with the name and position of the person you are sending it to showing through the window. Your business name and address should be in the upper-right corner. Make it look professional, because it is. Have the name and address of the person you are sending it to at the very top of the cover letter so it works as the salutation. Use a sans serif font in black, when you print it out, on a very white 50# paper. Finish with a thank-you, leave a space for your handwritten signature and type your signature, positioning your contact information below that. Sign it in blue ink. Print your news release on the same paper in black ink. Put it behind the cover letter, but fold them together, not separately, so the release doesn't get lost. Triple fold it so it fits perfectly in the envelope and crease the folds.

At the same time, you should be getting the word out to the beekeeping community, even a bit past the boundaries you will probably be serving. Use social media and newsletters to announce to the world that you will soon be here, that you will have these products and services available from these equipment suppliers and queen and package producers, and that you are here to help them with their beekeeping. Most may already know because you've been talking about this for months at the local club meetings you attend, but now it's official, and here are the wonderful details. Don't overdo this, but be relentless with your various social media messages. Stress who your suppliers will be, when you expect packages and queens to be available, the hours you will be open, and the classes you will have during the next season. Don't dump all this information at once, but release it in small amounts every few days on social media and every month in the newsletters. Send them to your website for even more information. That's at the top of your to-do list before you open—get that website up!

And then have a grand, grand opening. When you know you're ready, invite the world to your door. Make that invitation in social media and local new releases several times beforehand. Put out everything you can afford to have on display, have seasonal things readily available, and put special things—those only you will have—right out front with a good price. Hang a colorful sign telling the "NO FREIGHT" story. ("Come and get it and don't pay freight. NO FREIGHT!") Serve refreshments galore, maybe do a honey tasting of all the honeys you'll have, and if it's spring or summer, have some bees and equipment outside ready for show-and-tell. Get a few beekeepers from the local club to help with the bees, and let them show off a little. It's good for you and for the club. If possible, have your catalogue ready, but that may be a stretch this soon; at least have a price list ready to give away.

This is the "I need" customer . . .

YOUR CATALOGUE

Once folks know there's a new bee supply store nearby, they probably already have an idea of what's there. If they've been at this for any time, they only need to know what you don't have. And then they need the motivation to actually visit. Need is going to dictate some of that, when their smoker bellows disintegrates, they lose their hive tool, or it's time to get more supers. But, what about *want* rather than *need*?

You need a catalogue to get them in for wants. It's a budget item you can't ignore. You can't not have a catalogue. And it needs to be available before your customers need it. Your customers need to know about it, because you didn't have one last year, and they need to know how to get it. They need time to become familiar with you now that you are

no longer just a beekeeping friend, but have become someone they have to deal with, work with, and give money to. So try to have something ready before that grand opening. Leave it at the next meeting you go to, announce it on your social media, and promise that if they send an email, you'll send one out. Yes, you need a paper copy and a digital copy—on that website you already have up.

Catalogues are expensive, but they last all year, work for you every day, and don't need benefits, days off, or sick pay. You've looked at a thousand beekeeping equipment catalogues over the years, and you know the best and the worst you've ever had to work with. Make yours work like the best, and figure out why the worst was the worst. We're not going to build a catalogue here, but here are some of the

... and this is the "I want" customer.

things you should consider. Look at the catalogues you like and see how they accomplished these points.

- On the cover, run a photo of your store from the street or driveway so readers know what it looks like before they get there, and include as many of your employees as you can get. It may be only you and your bees and your spouse. That's okay.

- Put some teasers on the front like magazines on newsstands do: package day dates; field days; two cool new products; and the like. Add your logo, business name, address, and website, too.

- Just inside the cover, run a welcoming message, a story, a bit about being brand new and why you're glad to be here, what's new and different and why, and some more photos of your bees and your store, inside and outside.

- Part of this story tells about the people working with you. Feature somebody with a lot of bee-keeping experience, somebody with lotions and potions experience, somebody who contributes to one of the trade journals, or somebody who's an officer in a local group. Take advantage of that person, so to speak, and make sure customers know about their skills.

TIP: Resist the temptation to use as much space on a page for a $12 hand uncapping knife as a $2,500 uncapping machine. Consider the cost of an ad page's real estate. Certainly, you can use a photo of the knife—maybe even a photo of someone using it so there's no question about it. On the other hand, knowing your customers as well as you do, will you sell, maybe, one of the expensive uncapping machines all year and 150 knives. These need balance so your customer's needs are met immediately. A good guess is that most of your customers will drool over the uncapper, but buy the knife, right? This lends itself to at least a couple of pages of "absolutely necessary" stuff labeled as such. Smokers, uncapping knives, inexpensive veils, hive tools, and the like. Label the page and put small photos of smokers, three or four kinds, and the page they are on. Same with hive tools and all the rest. It makes surfing your catalogue easier for newbies and experienced beekeepers, alike, and once those newbies see what is necessary they are more likely to pick up more than they intended.

- Next up is a perfect table of contents. List what page everything is on. Don't skimp on this. Some customers, especially new customers, will devour it; some never read it. Have it, especially for those nonbeekeepers you are serving.

- Show a map of the area in about a 50-mile (80 km) radius featuring major roads to your store, and then a more-detailed map of your store's immediate vicinity.

- Put your phone number, hours, email, and website on every page, at the bottom, in a different colored ink.

- Say what's new. Lots of what you have is the same ol' same ol' that everybody has, but some things will be brand new on the market this year. You have to show that you're as good as everybody else, because you have what's new, too. And here's where you can get an upper hand: what do you have that nobody else does this year? What do you have that your customers can only get from you? That's a benefit and a problem solved. Work for that little gem. And don't forget to include those new items on the pages they would normally go with alongside the standards you've always had.

- Group your products by department: candle molds, supers, chemicals, protective gear, and so on. Keep each together and don't mix them up so customers can't find things. That said, sometimes you can put a new hive tool with the hive tools, but also with those new smokers, because they kind of go together and make a good impulse buy.

- When it comes to woodenware, put it all together, but really, really keep the different kinds distinct relative to quality, sizes, and types. Stress dimensions and sizes. This goes for the plastic and Styrofoam equipment, too.

- Every few pages, run a photo of a shelf with lots of your new and/or regular stuff on it. Don't put any label or identification; it's just a photo to show you have lots of whatever it is. It breaks up all the words and numbers and pictures of stuff you're talking about. It's just fun.

- In your section on protective gear, make it simple to get a bee suit that fits perfectly without needing to be tried on. Include diagrams for height, waist, chest, and length, and show where to measure for glove size.

- Dedicate a whole page to your custom extraction operation, so it looks and sounds simple and easy, and put it right in the middle of your honey harvesting pages.

- Show your bottles, jars, and labels. Show the jars and bottles together so the reader can get a feel for size comparison. Then, show collections of the categories—bears, queenline, skeps, and such. Make sure they all are full of honey, or honey-colored liquid, and they have lids on. Show labels actual size, and have something on one jar of each size so you know what it's going to look like. If you are doing custom labels, make the instructions easy to find and follow.

- Equipment and supplies for lotions and potions should be easy to find, and they should be all alone for those folks who will never buy beekeeping supplies. They should be right by your candle molds, wicks, ornaments, and gift pages. Group these pages toward the front so nonbeekeepers don't have to try and find them. You might tint the edge of the pages with a different color so they are really easy to find.

- Put feed and feeders together.

- If you have a complicated piece of equipment, show it in action, doing what it's supposed to do.

- If you're selling specialized equipment, for queen rearing for instance, put books on queen rearing in the same section.

- Explain drop shipping from the manufacturer carefully.

- Show the expensive equipment your supplier sells, too, to show you are in the same game and can handle anything a customer wants, no matter size of complexity.

- Run bee pages. They should include all about your packages, nucs, queens, and the quality and reputation of your supplier(s), plus put in an introductory how-to demo of package installation, queen introduction and anything else you can use to show you do know about bees and beekeeping. Solve problems, offer assistance, and know more than your customers.

- Books that cover specific activities or levels of skill should be with the equipment used in the book, plus on the pages with the other books.

- Tell us about the books you carry: title, author, length, size, cover, and a quote from a review. Or, get a quote from someone you know that has read it. Word of mouth goes a long way with a book. Don't assume that because you are selling them, customers will think they are good.

- If you are selling varietal honeys, put them in the catalogue, too. Tell a story about the more exotic kinds—when they will be available, or something about the beekeepers that produce them.

- When you are organizing photos for your catalogue, make them work for your website and other social media, too.

If you have something unique, like imported smokers, flaunt it in your catalog, store signs, and online.

WEBSITE AND SOCIAL MEDIA

The day you decide you might think about maybe going into this business is the same day you register your domain name. Don't put it off. Next, find a host for your page. It can be a local cable company or something much larger. Talk to all of them and tell them what you will be doing, and to make sure they can handle your kind and volume of business. You probably already have an internet service provider, but you'll probably need to upgrade that. And then develop a social media policy for your business and employees. There are websites to help with policy decisions; start there. They will stress what to do more than what not to do, but you do need some ground rules for employees or even family. Somebody will deal with your social media on a more or less steady basis; figure out who, where, and when. It's not too complicated, but not taking care of customers should never happen. And what about phones? You have one at the counter, probably one in the back shop, but any others? Most likely no, but that needs to be addressed.

If you have the time and experience, you might build your own website, but that's probably not the case. Like the catalogue, it's an expense, but it doesn't call in sick, need benefits, or take vacations. Hire someone to get it done. Get it done right the first time, making sure it can handle e-commerce, security, and everything else your attorney, insurance, and bank requires. At the same time, get your new email addresses situated. Get Mike@yourdomain.com, info@yourdomain.com, BeeSupplies@yourdomain.com, beekeeping@yourdomain.com, and maybe some others that point to you all registered, and set them all up to come to the same inbox. You can do this over time because you started early. You won't be selling anything soon, so e-commerce can wait, and you can add content as you develop it, but your social media campaigns start day one, so get all that geared up ASAP.

As early on as you can, begin sending out information. Make it seasonal, newsworthy, or photos or information people can use. Get a Facebook page for your business and put something out every day, or at least three times a week. Do the same with your twitter account, @yourdomain. Photos are a big part of this. Post them regularly, maybe as the new building is being built or your new beeyard is being developed. Initiate Instagram and Pinterest accounts so you have something on those sites, too. Photos can be used in all the platforms you participate in; some, certainly are photo heavy, others more word oriented.

If resources and direction permit, consider additional websites and Facebook pages. One will handle your beekeeping business, equipment, bees, and educational classes. The other will cater to your arts-and-crafts customers, your gift basket customers, and those who fancy your fancy honey. This is a totally different crowd, and you don't have to be the one talking to them. Consider someone who is teaching the classes you hold, or maybe someone in the family.

Take a class, and no, you won't be the oldest person in the class. Learning and using social media isn't a luxury or something you'll do when you have time later. If you are not part of this world yet, become one with your computer and mobile device. If that isn't in the works, hire your grandchildren, because they are. Your promotion and start-up budgets need to cover this, whether it's at a local extended learning center or through one of the many seminars training companies that do this full time.

You absolutely have to add your business to Google's business listings. Look at what others have done with theirs—not only businesses like yours, but also those in completely different fields. How do they approach new customers, handle service questions, and the like? Take a look at a lot of listings to figure out what will work best for you, and what your customers will want when they find this page, because they will. Have this ready some time before you open, so folks looking for bees or equipment will know when you are going to open, can sign up for emails from you, and will know when to visit. Don't be a surprise. Include a map, hours, phone number, website, and email; don't make it too cluttered, but don't skimp here. This is probably the first time your customers will meet you, and it needs to be as impressive as possible. You can pay to have this done and to keep it updated on a routine basis, but if you can handle your cell phone you can do this yourself. Change it occasionally, too. Different seasonal photos, new employees, package day, beeyards, and harvesting all work at the appropriate time of year. And announcements such as a bigger parking lot, expanded warehouse, being open more days or hours each day, or more colonies for pollinating all show that your business isn't static.

A BULLET-POINT TIP SHEET CAN WORK LIKE THIS.

(Be sure to use good photos.)

What to do in July in Northeast Ohio

- Do mite counts, if more than 1 mite per 300 bees, consider some form of control
- Before using a chemical control, make sure all honey to be removed is off
- One form of mite control is to make last season splits. July is the time
- Make sure you can get the genetics (queens) for splits this late in the season
- Local queens are always available at our store in July. Try some for your splits this season.

When it's time for your grand opening, take to Facebook. Post a lot of photos during the day showing the ribbon cutting, customers buying equipment, and the parking lot full of cars and trucks. Think about doing much of this as a Facebook Live event, well promoted before the opening and posted on your page as a video for later viewing. Link to it on your website, too. Assign somebody dedicated to this so it gets done right. Do five or six shows during the day, showing both inside and outside of your store and different sets of customers. Do some customer interviews on the live event. Have a celebrity present—a local beekeeping association president, for example, or a local commercial beekeeper. And have someone there who can talk about honey, arts and crafts, candles, baskets, and all the rest of the stuff you do at your business. Make sure to include everybody who will be a customer.

After the opening, continue to post photos and videos of events you sponsor, places you visit, displays, new equipment, unloading a big truck, or loading up someone's trailer. All signal you are doing a good job, business is good, and lots of beekeepers are benefiting. Using one of the many services available to bundle all your media blasts together means you only do this once, everybody gets to see and hear what's going on, and you spend a minimum of time actually posting all this.

You can also post interviews, both live and recorded. Hardly anybody does this form of advertising, which misses a great opportunity for word-of-mouth advertising. Find a good customer who is known in the local beekeeping community and do a short interview when they come to shop about why they visit, the type of service they get, the quality and selection of the products you sell, the convenience of the hours you are open, and how easy it is to get here—not all of these with one customer, but spread among several. Have them tell everything your operation does that helps them solve problems, save money, and make their life better. This word-of-mouth advertising is free, and it's the best there is. You can't buy it, but you can earn it. Then get that video shared and liked by everybody you can. Always, always add in your website address so people go to it for even more. In fact, you should have a sign right next to your front door and behind your counter with your website on it. Everything should have that address on it.

You can use your Facebook page, as is, for small advertising bursts, but the real value of Facebook advertising is the reach you can achieve. You can choose to reach beekeeping-interested users in your zip code, county, state, or surrounding states. For distant users, you can offer discounted freight or some other incentive for them to visit your website so you can further capture their attention. For those closer to you, offer whatever you can to entice them to visit, and to use your website so you can reach out to them again. Explore this avenue of reaching just the right people as much as possible. This is probably the least expensive method of reaching out, so learn how to use it to its full capacity.

 If you are selling equipment in more than one time zone, keep those distant customers in mind when you are scheduling your social media ads and posts so they don't get missed because they came out way too early in the day for some or too late at night for others.

You want as many people as possible to visit your website, and to make that happen you have to change it often. The landing page should always be familiar, but it should almost never be exactly the same; people will notice something's different, and you'll increase the probability that they'll look further. One way to do that is to have rotating stories, photos, or specials that stay on the screen for only a couple of minutes and then are replaced by something else. And once you have them on the page, give them something free for coming. What doesn't cost you anything that you can give away? Links to information, for starters. Share information you have vetted and that you stand behind. There is lots of trash available on the internet, and you want to be one of the sorters. Become the best place local beekeepers can go to for: safe, legal, and science-based how-to's; solid-backed political information that affects bees, beekeeping, and related causes; and anything you can find and support from a local, regional, or national association that your readers will need, will want, can use, and won't normally think about looking for. Make it easy for them to get this benefit and solve their problems, and they will keep coming back.

One of those sources of information should be you. If you can't put ten words together to make a sentence, that's okay. Put up a one- or two-page "What to Do This Month in Northeast Ohio" bullet-pointed tip sheet at least once a month, and send it to your local association newsletter editor so they can use it, too. Or put up a ten-photo story of you installing a package, a video tour of your beeyard in season, or a link to any of the videos you put on Facebook or YouTube. These posts should be personal, accurate, and not too long, maybe 10 minutes maximum. But you need to update them frequently, at least once a month. They promote you as the expert, and they solve a problem for anybody tuning in. And, because people don't know when you change these, offer to let them know. Provide a page for them to enter their email address. Every time you do something new, you'll let them know, and now you have an email with which to reach them whenever you want. Don't abuse that, but certainly don't ignore it, either. And be certain to have an unsubscribe option, too, if they want out of this.

If you have the time, you could also put out a free newsletter to those who sign up. Local association newsletters come and go, are good and bad, and can't always be relied on, and you can fill that gap. Feature announcements of meetings close enough for folks to go to, maybe for the several surrounding county groups they might not usually be in touch with, monthly how-to's, seasonal tasks, new equipment reviews, book reviews, and specials at your business. It takes time, but it's a vehicle that isn't used by many and can be beneficial to both you and your customers.

NEWSLETTERS, MAGAZINES, AND NEWSPAPERS

Speaking of newsletters, yes, you need to be noticed in your local and surrounding county newsletters every issue. You can approach this a couple of different ways, including an everything ad and a highlights ad.

An everything ad is a whole page that lists everything you do. Of course, you won't put every single item, but you'll note all of your catalogues sections. The association newsletter is where you first meet all those new beekeepers every year: the new ones from the classes last spring that didn't get a catalogue because you weren't open yet; those who moved into the county; and anybody who just found the club and joined. This kind of ad tells them everything they need to know about visiting you and what to expect, including location, website, email address, and phone number.

A highlights ad has a lot less information. It's nothing flashy or big, it's not very expensive, and color probably isn't in the works. It tells location, hours, website, email address, phone number, and that you are open for business and would love to see them. Change it a couple or three times a year. You can even use the same content but a different shape—vertical as opposed to square or horizontal. You should consider a one-time new season ad, naming all the new equipment you will have this year, your extended hours, and when package day will be. Then, mid-season, mention boxes for honey, hive stands, and nuc boxes. Finish in the fall with bottles

and labels. Time your news to just before your customers need it.

Do the same for the state association newsletter—a small ad, but always there, with maybe one or two larger splashes seasonally, especially if you are the only operation that does something nobody else in the state does, such as getting your packages from the best producer in northern California, or the best harvest equipment from Italy.

National magazines? That depends on who your customers are, and maybe who your suppliers are. If you have migratory customers in the summer who spend the rest of the season five states away, or who show up only for a single crop to pollinate, you want them to know you exist for any emergency that comes up, and that there's a local source for anything they might happen to need.

Local newspaper ads are expensive and reach a lot of people who won't read the ad once they notice something about bees. So don't advertise about bees. Promote your lotions and potions, classes, gift baskets, honey, pollen, and propolis. Include a "Curious About Keeping Bees?" line in there somewhere announcing the beekeeping class coming up in the spring.

Local shoppers are less expensive, and they're usually less productive. But you should try them once or twice to see, if your budget allows. Direct mail is the most expensive advertisement per reader reached, but if you have access to mailing lists that will reach select customers or neighborhoods, it may do well for you.

You should also consider co-mailing or co-advertising prospects. Advertising together with a local nursery may be beneficial, because the folks reading that ad are already predisposed to the natural world. Putting your honey in a local restaurant ad might work well, especially if the restaurant has a recipe with your honey in it. Putting your lotions and potions in a beauty salon ad should work well. With any of these you get additional, more focused exposure for your money.

Advertising is often the first budget to get cut when finances are an issue, and that's understandable. Someone once said, "I know only half my advertisements pay off, but I don't know which half," and that's often the case. It's a gample far too often, which is why you have to work at measuring the effects of every ad you have to pay for. Promotion codes for websites, Facebook, and other social media are one good way to see if these platforms are working for you. Measuring visits to your website after placing one of the other types of ads helps, too. The number of followers you have on social media platforms is another metric to use, especially when you can figure out what triggered an increase. In any event, don't shoot in the dark. Try and measure, then try again. Find the best social media platform for you and concentrate there, and spend the least amount of money on purchased ads you can to get the best results.

 If you haven't already, create a YouTube channel for your business. And then use it often. Your videos don't have to be long, fancy, or perfect. Show a new product, a just-learned technique, introduce a new employee, or talk about the upcoming candle class or other events taking place this week either at your store or elsewhere locally and/or through a trade association. Send out your social media ads announcing your new video with a link to the video. It's simple and it's personal.

KILLING A BUSINESS IN TEN EASY STEPS

1. Don't Advertise! Just pretend everybody knows where you are, when you are open, and what your website address is.

2. Don't Advertise! Tell yourself you just don't have time to spend thinking about promoting your business or learning all that new social media stuff.

3. Don't Advertise! Just assume everybody knows what you sell.

4. Don't Advertise! Convince yourself that you've been in business so long customers will automatically come to you.

5. Don't Advertise! Forget that there are new potential customers who would do business with you if they were reminded and urged to do so.

6. Don't Advertise! Forget that you have real competition from the internet and other stores who attract customers all the time.

7. Don't Advertise! Tell yourself it costs too much to advertise, and that you don't get enough out of it.

8. Don't Advertise! Overlook the fact that advertising is an investment in selling—not an expense.

9. Don't Advertise! Be sure not to provide an adequate advertising budget for your business.

10. Don't advertise! Forget that you have to keep reminding your established customers that you appreciate their business.

5 Staying in Business— or Not

Measure *everything* you do and what it does for you. Maybe you started this adventure because it just sort of happened. You had enough bees to make enough honey to sell in enough places that you made enough money that you thought you could see your way to quitting your day job and doing what you really wanted to do. And if just keeping more bees is all this is about, then you don't have many things to measure. Maybe improving your labels, using more and different containers, adding more varietals, trying some infused honeys, selling more and different honey sticks, and branching out just a tad into creamed and comb honey is about all you'll have to deal with. But you still have to measure. Are you making a good enough return on the varietals? How much to redo the labels for everything you have? How many pallets of bottles can you afford, and do you have room for them? Do you get stuck with crystalized comb honey too often? Have you been able to find new outlets that are close enough to be economical, yet far enough away so as not to siphon off sales from existing accounts? And this is just for honey and bees.

Or, it may have gone the other way. You had bees and honey, but you were in a position to expand that a lot by buying out an existing small supplier who wanted to retire. Then you had honey, bees, equip-ment, and wax, and you wanted to expand all of that to a bigger and better selection. So fairly quickly, the honey part of this took a back seat, and you had a store and a cash register and a catalogue. Now the measuring gets complicated, though possible. Then an employee comes along, and insurance goes up.

Your goal is to find the right balance of what you spend your time doing, and that means making every minute you spend on the job make the most money possible. Ask these questions: Is honey still the best way to go, or do you still have bees because that's what you like to do best? Is selling packages worth the chaos those few days cause every spring? Are the headaches with mismatched woodenware worth the few dollars saved? Are you ready to do something very bad to the next teacher who brings in a bunch of loud, troublemaking kids? Once you've been at this a bit, you have to measure all of this to decide how to proceed.

Find the Right Balance

And, at the same time, what haven't you done that maybe you should have to increase your income per hour? What can you begin that won't eat a ton of time, but will make a decent return for you and not compete with everything else you are doing? What are you doing that you could be doing a lot more of? What could you stop doing without noticing a change in income?

After only a season, two at most, you will likely have a good feel for much of this, but the smart oper-ator doesn't have to wait a season or two. They'll see it before they open the door the first day. The trick to making good observations is to make good measure-ments, starting before you open the door. When you sent out that first press release, what were the first five questions somebody asked at the next association meeting you went to? What questions came next?

How long do you want to keep making candles?

Diversity of Revenue

Successful businesses get this right quickly. They choose a model that works and already exists and use it right off, or they measure fast and make appropriate changes as soon as it's apparent changes are needed. A small business, one without a ton of resources or experience, one without a lot of business models to study, can waste a lot of time and money trying to find that balance. The right mix of goods and services is going to be different for every bee supply operation on the planet. Maybe not a lot different—most of your customers just need to know what you don't have—but still it'll be different. Employees can be hard to find or to keep, the location may be too far from your customer base, a competitor may move in, or you may have three crop failures in a row. These obstacles need to be overcome to make the business work, but the right mix in your revenue stream should be able to absorb any of them in a given year. Enough of this, some of that, just a tad of something else keeps the ball rolling and your income, if not high, at least steady. Measure.

You would be surprised at how many suppliers evolve into gifts, gadgets, and collectibles business . . .

Quality of Life

But that income stream may not be what you want to do. It's common to start something the way you think it should be, then watch it morph into something entirely different—something you don't want to spend your time doing. Maybe you're not good with kids, but the local school really likes sending them out to visit because the principal is a local beekeeper and feels the experience is just what children need early on, and you're the only person in town that can make that happen. The money is nothing to laugh at, and the parents keep coming back as customers, and there are new ones every year. And so it goes. You end up being a branch of the local school, and it just gets better and more lucrative, but it's not what you wanted to do. Finding someone to do that part of the business is one way around this, and though there's a cost, it can keep things going.

This is a tough decision. Do you change the rules and get back to doing what you wanted to do in the beginning, or at least rearrange the focus of what you do every day? Do you keep doing what you don't want to do but are making a living at? Isn't this why you quit your day job in the first place? If you're not happy doing what you're doing, what do you do instead?

Another quality-of-life problem lies in the other direction. You're working 12 or 15 hours a day just to keep the doors open, and it's a stretch to make ends meet every month. It seems Amazon really is as bad as you were told. Your family doesn't see you nearly as often as they used to, and you have to think twice about turning the heat up because of the cost. You're not working for the man any more, but the man took better care of you when you were. Do you give up, refocus the revenue streams, or what?

There are more obstacles, certainly. The trick is to recognize them before they become obstacles so you can make changes sooner, and so you can make better decisions based on facts gleaned from good measurements. Successful businesses don't just happen. They stay successful because of good planning and hard work. I've already said it, but "Build it and they will come" is a pipe dream. What isn't a dream is "Make good plans, build it, and work hard to keep them coming." That's what makes a success.

FOCUSING ON FEWER ASPECTS

You need to fix your imbalance before it breaks you. Go back to why you started this in the first place. What were you doing that got you to the point that you saw a way to make it full time? Are you a beekeeper at heart, able to make honey in a parking lot, keep lots of bees healthy no matter what, raise queens, make splits, get bees moved anywhere anytime, and figure out how to fix a truck transmission in the middle of the night in the middle of nowhere? You had more places that wanted your honey, wanted your queens, wanted your bees, and wanted your teaching skills than you could possibly serve while working 40 hours a week for someone else. But to make it a full-time job, you needed more honey, bees, and pollination contracts than you had. So you branched out. A few things here, a few there, a deal with a manufacturer for the rest, and suddenly

. . . If it works, don't fight it.

you're a bee supply store with store hours, inventory, advertising and catalogues, school tours, employees, and all the rest. You can imagine how often that happens and how often it fails.

Or, were you a get-by beekeeper, kind of having fun, and it got a little carried away? But you found you could, indeed, sell anybody honey, and you found honey for sale from beekeepers who hated selling honey. Then you found dozens of places that wanted to sell your honey for you, and paid a good price to do just that. And you could get stuff for beekeepers from the strangest places and for almost nothing and make money selling that, too. You were able to see dollars where others saw only pennies, and you did it with ease and grace. But you couldn't get bigger while keeping your day job. So, you branched out a little. Then a lot. Then the store got in the way of the honey. And suddenly things fall apart. You've got store hours, inventory, advertising and catalogues, school tours, employees, and all the rest. And bees come in third all the time. You can imagine how often this happens, too.

So, refocus. Which of these scenarios is why you quit your job in the first place? Do you want to keep bees with a little bit of retailing equipment, or do you want to run a store and have a few bees out back? Choose one, before it all goes to pieces.

> **TIP:** The only way to figure out which side of this coin you're on is to measure, measure, measure. Which part of the business makes the most money? Is it supporting the rest of the business? Which part do you hate to have to do and which do you love to get up for in the morning? Not only that, what part of this—no matter the joy of the job—do you spend the least time on and make the most money on? Measure, measure, measure.

For some, bigger is always better.

Too Much Success

Yes, it happens. Be prepared. When you started this adventure, one of your options was to take over a bee supply business that was for sale—buy the property, the inventory, the accounts, and the rest and not miss a day's work. That can work the other way, too. Out of the blue, someone may walk in one morning, look around a bit, and ask, "How much?" It may be someone local who knows the value and scope of the business you are running, or someone who sees the potential of what you are doing where you are doing it. Either way, how much?

Imagine the questions you immediately need to know the answers to the following:

- Is anybody in my family interested or capable of taking over this operation?

- Do I want to quit doing this?

- Do I want to quit doing some of this?

- What is this operation worth, anyway?

- What is just the inventory worth?

- What about just the honey accounts?

- What is the land I'm sitting on worth?

- Hives, too?

- And a hundred more questions.

That sort of instant-retirement situation doesn't come along very often, so if it does be very careful, and get your money's worth. What do come along are opportunities to prepare for retirement from day one. It absolutely has to be in your overall plans. When you were working nine to five, you had personal savings, an automatic 401(k) deduction, maybe a pension, your spouse's income, perhaps, and other investments. But when you left, you left much of that behind, and those retirement plans were no longer automatic. Now you have to do it yourself.

Start by determining what you'll need to retire and what you'll need to do to get it. Don't forget to factor in something for what inflation will do ten years down the road. Go online and find an investment calculator, offered by most major financial institutions. They will give you insight into what you need, what you need to save, and the increase in value from making that money work for you.

What will you do when you retire? That's an important question. Will you continue to work, say by keeping bees? Or will you be a wild spendthrift, finally buying that cabin in the mountains, and visiting all of Europe for a month each year? One gets you money and spends, probably, less than you are living on now. The other is a sure way to the poorhouse if you don't prepare for it.

Another unknown is how long you and your spouse will live. A man turning 65 in 2018, according to the Social Security Administration, will live to on average 84, a woman to 87. One in four of those will live to 90. So will you run out of money before you're gone? It's an interesting, kind of morbid, question, but one that needs to be studied. So start saving right away. If you can, set aside at least 10 to 15 percent of your annual income. It's easy to promise yourself that you'll save more next year, but this year it's tight and you need the money more than you will twenty years from now. Don't.

Start with your banker, who has boatloads of information on retirement and estate planning. You can probably set up an automatic deposit, much like a 401(k), so you don't have to think about it. You won't know until you ask. Every major financial institution would like your business, too, so go online and check them out. It may be easier or cheaper or they may do better with investing than your local bank, but check out all of your options. The government wants to help, too. The IRS and Social Security both have planning tools on their websites. And don't overlook local colleges and tech schools that offer classes and seminars on estate planning for business owners. Check online for extension bulletins, and look again at the list of resources you collected when making a business plan. You can pay someone to do this for you, too. Financial advisors are everywhere, and for a fee they will arrange all this for you.

And when you've got all this under control and that guy in the flashy blue suit walks in, looks around, and asks, "How much?" you'll have the right answer, and *In Business with Bees* will have completed its course.

Congratulations!

Acknowledgments

Lots of people have to come together to make a work like this happen: publishers, editors, fact checkers, proofreaders, designers, layout specialists, printers, distributors, and booksellers. I absolutely appreciate every one of these people, and those I've unintentionally overlooked, because this is what I do every day. I get to make books and magazines materialize out of good ideas from the authors who write for our magazine, and the good people who take those ideas and get them on paper, and get them into the mail and the hands of readers who want, and need, what we do. So I thank all of you who also get to do this. We are fortunate, and I am most fortunate for what each of you do, every day.

Of course there are special people who have helped with this from the other side. Four people, among many, who have shared what they do, the mistakes they have made, and the good choices they have made to make their businesses a success. I refer to Denzil and Sheila of Queenright Colonies, an amazing operation here in northeast Ohio that started more bees than store, and ended up more store than bees—an excellent model for that direction—and to Dan and Bonita Conlon of Warm Colors Apiaries in Deerfield, Massachusetts, which is more bees than store, and is an excellent example of that side of the coin. Thanks for sharing what you know and what you've done, good people.

About the Author

Kim Flottum has been the Editor of *Bee Culture* magazine for more than 30 years. From that perch, he's been able to visit and get to know scores of beekeepers who have gone from beginner to backyarder, then to sideliner and, finally, to running a beekeeping business. He's watched many grow and prosper for as many reasons as there are businesses, and watched many fall by the wayside due to bad luck or bad choices. By running a magazine, he's become keenly aware of those beekeeping businesses that advertise in that magazine. And after 30 years of watching, he has uncovered those that are successful, and more importantly why, and those that are not, and just as importantly, why not. This book is the essence of what he has seen and learned in that time, from those people.

Index

Also available from Quarry Books!

The Backyard Beekeeper
4th Edition
978-1-63159-332-1

The Backyard Beekeeper's
Honey Handbook
978-1-59253-474-6

The Beekeeper's Journal
978-1-59253-887-4

The Beekeeper's
Problem Solver
978-1-63159-035-1

The Benevolent Bee
978-1-63159-286-7